Behaviormetrics: Quantitative Approaches to Human Behavior

Volume 21

Series Editor

Akinori Okada, Professor Emeritus, Rikkyo University, Tokyo, Japan

This series covers in their entirety the elements of behaviormetrics, a term that encompasses all quantitative approaches of research to disclose and understand human behavior in the broadest sense. The term includes the concept, theory, model, algorithm, method, and application of quantitative approaches from theoretical or conceptual studies to empirical or practical application studies to comprehend human behavior. The Behaviormetrics series deals with a wide range of topics of data analysis and of developing new models, algorithms, and methods to analyze these data.

The characteristics featured in the series have four aspects. The first is the variety of the methods utilized in data analysis and a newly developed method that includes not only standard or general statistical methods or psychometric methods traditionally used in data analysis, but also includes cluster analysis, multidimensional scaling, machine learning, corresponding analysis, biplot, network analysis and graph theory, conjoint measurement, biclustering, visualization, and data and web mining. The second aspect is the variety of types of data including ranking, categorical, preference, functional, angle, contextual, nominal, multi-mode multi-way, contextual, continuous, discrete, high-dimensional, and sparse data. The third comprises the varied procedures by which the data are collected: by survey, experiment, sensor devices, and purchase records, and other means. The fourth aspect of the Behaviormetrics series is the diversity of fields from which the data are derived, including marketing and consumer behavior, sociology, psychology, education, archaeology, medicine, economics, political and policy science, cognitive science, public administration, pharmacy, engineering, urban planning, agriculture and forestry science, and brain science.

In essence, the purpose of this series is to describe the new horizons opening up in behaviormetrics — approaches to understanding and disclosing human behaviors both in the analyses of diverse data by a wide range of methods and in the development of new methods to analyze these data.

Minoru Nakayama • Wioletta Nowak

Development of Diagnostic Procedures Using Features of Pupil Reactions

Springer

Minoru Nakayama
Department of Information and Communications Engineering
Institute of Science Tokyo
Meguro-ku, Tokyo, Japan

Wioletta Nowak
Department of Biomedical Engineering
Wroclaw University of Science and Technology
Wroclaw, Poland

ISSN 2524-4027 ISSN 2524-4035 (electronic)
Behaviormetrics: Quantitative Approaches to Human Behavior
ISBN 978-981-95-4501-8 ISBN 978-981-95-4502-5 (eBook)
https://doi.org/10.1007/978-981-95-4502-5

This Springer imprint is published by the registered company Springer Nature Singapore Pte Ltd.
The registered company address is: 152 Beach Road, #21-01/04 Gateway East, Singapore 189721, Singapore

Preface

This monograph book summarises some of the challenges of developing diagnostic procedures using pupillary changes and is a work in progress.

Both editors were originally interested in the dynamics of pupillary change, and physiological or psychological factors were the main focus of their research. After discovering intrinsically photosensitive retinal ganglion cells (ipRGC), the excitement of the pupil research groups has led to new research results once this new paradigm on pupillary response had been examined. By synthesising this with conventional research works, the possibilities of these new approaches to research and new applications to diagnosis are discussed.

Though the editors could not follow the movement of all new research activities, we were fortunate to have Prof. Satoshi Ishikawa, MD, bring us up to date on what is happening in this area of research. In addition, he has warmly supported the collaboration with Prof. Hitoshi Ishikawa, MD's research group at Kitasato University in Japan. The group already had a number of Age-Related Macular Degeneration (AMD) patients and had already observed pupillary changes in the pupil light reflex (PLR). During this collaboration, we learned about the physiological background of patients and about issues regarding diagnosis. This experience became the trigger to the study of this topic.

One of editors works at the university in Wrocław, Poland, where Prof. Aloysius Alzheimer, MD, discovered the disease of dementia, currently known as "Alzheimer's Disease". In those days, the Wrocław Medical University, where the first Alzheimer disease laboratory was established, was known as the University of Breslau. Due to a collaboration between the editors of this manuscript, some clinical survey data can be shared between the members of the various labs involved in the research into developing new diagnostic procedures. Generally, clinical surveys are based on a limited number of participants, so the development of a technical diagnosis procedure is limited by this. Fortunately, an additional set of survey data was obtained independently by the Japanese researchers. Prof. Masatoshi Takeda, MD, supported clinical survey work at a series of local clinics. Several new diagnostic procedures were introduced step by step.

Here, the editors would like to present our deep appreciation for all the support and collaboration of many experts who participated in this research work. The remaining issue is the decoding of pupilograms of participant's responses and classification of patients using technical approaches such as data science.

However, the main issue of developing a prediction procedure for diagnosis has not yet been resolved sufficiently, which is unfortunate. Once again, this monograph introduces some new diagnostic procedures, but their possibilities are still being examined. Therefore, the editors would like to ask our readers to contribute their ideas and offer suggestions. We are looking forward to receiving your contributions.

Meguro-ku, Tokyo, Japan
Wrocław, Poland
January 16, 2025

Minoru Nakayama
Wioletta Nowak

Acknowledgements The editors would like to thank Dr. Akinori Okada, the editorial supervisor of the series of the monographic series: Behaviourmetrics: Quantitative Approaches to Human Behavior, who invited us to summarise this monograph. Also, we wish to thank Yutaka Hirachi, of the editorial management division of Springer Japan.

Once again, the editors would like to thank the co-operators for the chapters they have contributed, as follows: Prof. Hitoshi Ishikawa, MD, Prof. Ken Asakawa, MD. Yoshiaki Ichibe, MD. Tomasz Kręcicki, Prof. Elżbieta Trypka, MD, MD. Artur Andrzejak, Dr. Andrzej Hachoł, Dr. Anna Żarowska and all of the collaborators of the experiments.

Additionally, anonymous reviewer provided kindly significant comments and suggestions to shape all chapters. The editors appreciate the contribution.

The editors also wish to thank all of the publishers and journal organisations which agreed to the republication of the manuscripts, such as Springer & Nature, the European Alliance for Innovation (EAI), the IEEE and the Association of Computing Machinery (ACM).

Contents

Chapter 1
Extracting Irregular Pupil Light Responses to Chromatic Stimuli Using Waveform Shapes of Pupillograms

Abstract The waveforms of the Pupillary Light Reflex (PLR) can be analysed in a diagnostic test that allows for differentiation between disorders affecting photoreceptors and disorders affecting retinal ganglion cells, using various signal processing techniques. This procedure has been used on both healthy subjects and patients with Age-Related Macular Degeneration (AMD), as a simple diagnostic procedure is required for diagnosis. The Fourier descriptor technique is used to extract the features of PLR waveform shapes of pupillograms and their amplitudes. To detect those patients affected by AMD using the extracted features, Multi-Dimensional Scaling (MDS) and clustering techniques were used to emphasise stimuli and subject differences. The detection performance of AMD using the features and the MDS technique shows only a qualitative tendency, however. To evaluate the detection performance quantitatively, a set of combined features was created to evaluate characteristics of the PLR waveform shapes in detail. Classification performance was compared across three categories (AMD patients, aged and healthy subjects) using the Random Forest method, and weighted values were optimised using variations of the classification error rates. The results show that the error rates for healthy pupils and AMD-affected pupils were low when the value of the coefficient for a combination of PLR amplitudes and features of waveforms was optimised as 1.5. However, the error rates for patients with age-affected eyes were not low. A classification procedure for AMD patients has been developed using the features of PLR waveform shapes and their amplitudes. The results show that the error rates for healthy PLRs and AMD PLRs were low when the Random Forest method was used to produce the classification. The classification of pupils of patients with age-affected eyes should be carefully considered in order to produce optimum results.

Chapter Co-authors: Minoru Nakayama, Wioletta Nowak, Hitoshi Ishikawa, Ken Asakawa, and Yoshiaki Ichibe

M. Nakayama, W. Nowak, *Development of Diagnostic Procedures Using Features of Pupil Reactions*, Behaviormetrics: Quantitative Approaches to Human Behavior 21, https://doi.org/10.1007/978-981-95-4502-5_1

1.1 Introduction

The pupillary response has long been used for diagnostic procedures [14] and psycho-physiological studies [3, 12, 21]. The pupillary response, a reaction to light intensity, is well known, and the pupillary light response controls the dilation and constriction of the pupil in response to changes in light intensity. The response is usually observed and recorded using pupillograms which consist of pupil diameter and time diagrams [40].

This response is often called the pupillary light reflex (PLR). It has also been used as an objective measure of retinal and optic nerve functions. In recent years a growth in interest in the examination of PLR has been observed. It is the result of the discovery of a new type of retinal ganglion cells [5, 11]. Since the melanopsin-associated photoreceptive system (ipRGCs: intrinsically photosensitive retinal ganglion cells) in the human retina was discovered [9], various diagnostic procedures have been proposed and introduced.

In particular, the difference between the response behaviour of this photoreceptive system to some types of chromatic stimuli and the behaviour of the conventional rod-cone system has been compared [9] to determine the possibility of detecting the condition of the retina. In addition to receiving rod and cone inputs, the responses based on the newly discovered system have been studied as intrinsically photosensitive [19, 41]. Those cells are interchangeably referred to as intrinsically photosensitive retinal ganglion cells (ipRGCs) or melanopsin-mediated retinal ganglion cells (mRGCs), and according to recent studies they drive pupillary responses and circadian rhythms [5, 9, 11]. Also, this phenomenon has been applied to diagnostic procedures for patients with glaucoma [7] or retinitis pigmentosa [17, 18]. Feature extraction and analysis of PLRs has made a significant contribution to the development of diagnostic procedures for these patients, as opthalmologically scientific evidence was discovered. Another well-known disease that is observed in ageing patients is Age-Related Macular Degeneration (AMD) [39]. This disease is related to the condition of the retina, and thus PLRs are affected by the progress of the disease. Unfortunately, this disease spreads gradually. Therefore, detection and prediction procedures are required. A simple procedure that detects prominent symptoms, such as an easy at home test, is necessary.

In addition to some types of portable pupillometers currently available, a PC web camera and a smart phone with an additional lens can be used to observe pupillography. Some types of clinical consultations can be conducted anywhere using either of these pieces of equipment.

As mentioned above, PLRs can be applied to extract features and to detect irregular responses. This means that a record of temporal pupillary changes as pupillograms is a time series signal, requiring various types of signal processing

Originally published as "Discovering Irregular Pupil Light Responses to Chromatic Stimuli Using Waveform Shapes Blue of Pupillograms", EURASIP Journal on Bioinformatics and System Biology, # 18, pp. 1–14, October 2014.

techniques. This suggests that a simple signal processing technique may be used to extract some symptoms of the disease. For example, the Fourier descriptor technique is often used to indicate the shapes of waveforms [32, 42, 43], and the possibility of using this method should be examined.

In this chapter, feature expressions of PLR waveform shapes of pupillograms are introduced to identify pupil characteristics. Also, the possibility of detecting eyes that are affected by AMD and other factors is examined. Therefore, the following topics are addressed:

1. A procedure for extracting features of PLR waveforms of pupillograms is created, and the features used to compare these waveforms are analysed to detect irregular responses.
2. To determine the possibility of detection in diseased eyes, several classifying techniques were applied to features of PLRs, and the performance of the techniques is discussed.

1.2 Related Works

1.2.1 Characteristics of ipRGCs

The study of intrinsically photosensitive retinal ganglion cell (ipRGC) mentioned in related research work currently being conducted is concerned mainly with, on the one hand, the study of the ipRGC structure and functions and on the other hand, with the diagnostic use of their specific activity. A brief review of literature regarding this research is presented below.

The ipRGCs are atypical retinal photoreceptors distinct from classical rod and cone photoreceptors [36]. They express the photopigment melanopsin and are intrinsically photosensitive, since they showed sluggish melanopsin-mediated responses. They can also act as conventional RGCs by receiving synaptic rod/cone input via bipolar cells. This integrated information is then transmitted to numerous discrete brain regions involved in both non-image and image-forming vision [35]. The ipRGCs ultimately modulate a multiplicity of behaviours including circadian photoentrainment, PLR, activity masking, sleep/arousal, anxiety and light aversion and even make a significant contribution to visual functions. Recently, it has been discovered that ipRGCs consist of several subtypes that are morphologically and physiologically distinct, which contribute differentially to the above-mentioned non-image and image-forming functions [10, 35]. A detailed study of the ipRGC's different types and their behaviour with varying attributes of light is being still analysed. In particular, intensive research involving the relationship between melanopsin activity and PLR reflex is now being conducted [8, 31]. These studies have direct applications in clinical conditions for the diagnosis of retinal degeneration and sleep disturbances under clinical conditions [8, 25].

The application of ipRGCs to the study of retinal degeneration is focused mainly on the use of pupil responses to chromatic light as a clinical marker, to allow differentiation between disorders affecting rod/cone photoreceptors (the outer retina) and those affecting retinal ganglion cells (the inner retina). As used above, the terms "inner retina" and "outer retina" are the consequence of the anatomical distribution of ipRGC photoreceptors (the inner nuclear layer) and rod/cone photoreceptors (the outer retinal layer) in the retina [22]. Where a disease affects multiple retinal layers, the pupillary light reflex could be a useful tool in determining the contributions of the inner and outer retina to the disease process.

Previous studies suggest that the influence of ipRGCs on features of PLR in response to chromatic light stimuli has been discussed, such as the reaction behaviour of the pupil. Specific behavioural characteristics will be summarised, as follows.

1.2.2 PLR for Chromatic Stimuli

Kankipati et al. focused on PIPR (post-illumination pupil response) analysis of glaucoma patients and compared them with normal subjects [15, 16]. They used a 10 second light stimulus (retinal irradiance: 13 log quanta/cm^2/s, light wavelengths: 470 nm (blue) and 623 nm (red)) and recorded pupillary response for 50 seconds after light cessation. They found that normal subjects displayed a significant PIPR for blue light (but not for red light), which is consistent with the proposed melanopsin-mediated response. When glaucoma patients are compared to patients using age-matching controls, there was a significant decrease in melanopsin-mediated PIPR. They concluded that PIPR has the potential for use as a clinical tool for evaluating patients with glaucoma. Feigl et al. [7] also tested PIPR to analyse whether glaucoma alters the function of ipRGC. They use a 10 second light stimuli with 488 nm and 610 nm light wavelengths and retinal irradiance of 14.2 log photons cm^2/s. They have found that patients with advanced glaucoma have a dysfunctional ipRGC-mediated PIPR. It has been confirmed that PIPR may be a clinical indicator of progressive changes in glaucoma.

Kardon et al. focused on the percentage of pupil contraction in transient and sustained pupil responses in patients with retinas pigmentosa [17, 18]. They used a stimulus paradigm using red and blue light as a continuous Ganzfeld stimulus which produced a 13 second stepwise increase in intensity over a 2-log-unit range (low (1 cd/m^2), medium (10 cd/m^2) and high (100 cd/m^2)). They have found that pupil responses to red and blue light stimuli which are weighed to favour cone or rod input are significantly reduced in patients with retinas pigmentosa. Their preliminary results suggest that pupil response to a low-intensity blue light, a high-intensity red light and a high-intensity blue light may be reasonable markers of rod activity, cone-driven responses and direct, intrinsic activation of mRGC, respectively.

Specific features of PLRs such as PIPR are observed in response to intensity, pulse duration and wavelengths of light according to colour. Also, reactions of PLRs are different with activated cone-driven or ipRGC-based ganglion cells. As the features of PLRs provide significant information about the detection of AMD, the diagnostic procedures have been investigated.

1.2.3 *Influence of AMD*

Age-Related Macular Degeneration (AMD) is another concern regarding retinal disease. This disease causes impairment of both the inner and outer retinal layers, depending on the stage of the disease, and pupil chromatic response measurements may allow the monitoring of the progression of the disease or facilitate the determination of different stages of the disease [6]. In particular, the function of ipRGC concerns the PLR, while AMD influences the waveforms of PLR [8].

Brozou et al. tested outer retinal contributions to pupil responses in patients with age-macular degeneration (AMD) [4]. The study showed that AMD significantly affects the pupil's response to light stimulus (20 msec duration and 24.6 cd/m^2 intensity) when compared to normal subjects. Feigl and Zele present a new experimental paradigm for the first time, which allows the differentiation of inner and outer retinal contributions to the pupil response in AMD [8]. They used a 11.9 sec duration 0.5 Hz sine wave stimulus with 464 nm and 635 nm light wavelengths and retinal irradiance of 15.1 log photon $cm^{-2}s^{-1}$. Additionally, they introduced a new metric called "a phase amplitude percentage" (PAP), which reflects inner and outer interactions. PAP is determined from the average long-wavelength and short-wavelength peak-to-trough phase amplitudes. PAP approaches zero for retinal irradiances below a certain melanopsin threshold, as PLR is driven by rods and cones. PAP is nonzero for retinal irradiances above a certain melanopsin threshold, as PLR is predominantly driven by cones with ipRGC contributions. They also showed that application of this paradigm to AMD patients can provide information that ipRGCs are altered. This result can be a very important step towards using PLR to determine the inner and outer retinal dysfunction of AMD patients. Future work will be focused on quantifying retinal inputs to the pupil response in order to determine the different stages of AMD or to monitor the disease's progression.

AMD affects some parts of the retinal condition as retinal dysfunction; their retinal regional positions influence relatively on responses of PLRs in coordination with cone-driven responses. In most clinical observations, these influences appeared in a peripheral region of the retina where patients are unable to recognise symptoms in the early stages. Also, the condition can happen to either eye. The empirical evidences should be considered in order to create a better diagnostic procedure.

1.2.4 Detecting Symptoms of AMD

Our work presents an attempt to use the pupil response to chromatic rectangular light pulses, which are routinely measured in clinical studies, as a simple and fast tool (indicator) for early detection of Age-Related Macular Degeneration (AMD). AMD (both the dry and wet forms) is the one of the most common irreversible causes of severe loss of vision. It usually affects older adults and results in a loss of vision because of damage to the central part of the retina, known as the macula. The gradual disappearance of the retina pigment epithelium (RPE) in the dry form results in patches of chorioretinal atrophy lacking any visual function. In the wet form, the damage is due to the escape of subretinal fluid/intraretinal fluid, blood, or destruction of photoreceptors and RPE by fibrous or fibrovascular tissue. Since AMD destroys photoreceptors that are initial receptors for PLR, it may reduce the PLR. The type or localisation of the damaged photoreceptors may then be indicated using a pupil response to chromatic light. AMD is painless, and, consequently, the lack of early warning signals in existing retinal pathology necessitates an indicator which will allow for earlier and easier detection (prediction) of this type of disorder. It is believed that the pupil reaction to chromatic light could be such an indicator.

Some empirical clinical observations may suggest that some symptoms influence the waveform shapes of PLRs in response to chromatic light pulses where the light intensity, duration and wavelength are controlled. The representations of the waveform shapes of PLRs may be a clue to correct diagnosis of the patient. For measurement taking, a number of conventional characterising techniques are introduced, as follows.

1.3 Methods

1.3.1 Subjects

Six healthy young subjects (20–21 years old) and six elderly AMD-affected patients (59–86 years old) participated. Each elderly patient had a diseased eye and a normal eye, according to a medical doctor. In this work, three categories of eye condition are defined as follows: Disease eye (D) is patient eye, another normal eye of elderly patient is normal (N) and eyes of young healthy participants are healthy (H).

Diseased eyes with choroidal neovascularization (CNV) in the macular region are often affected by new blood vessels, which bleed and form dense macular scars [39]. Also, CNV is a major cause of visual loss due to AMD [39].

Patient details are summarised in Table 1.1. Regarding fluorescein angiographic assessment as a medical diagnostic [39], the diseased eyes are classified into *predominantly classic CNV lesion* (the area of classic CNV occupies 50% or more of the entire lesion) and *occult lesion* (either fibrovascular pigment epithelial detachments or late leakage from an undetermined source [30]). In regard to the

Table 1.1 Patient information

	Age	Sex	Type of disease	Visual Acuity D	Visual Acuity N
Patient 1	82	Male	Predominantly	0.2	1.0
Patient 2	66	Male	Predominantly	0.04	1.0
Patient 3	74	Male	Predominantly	0.4	0.9
Patient 4	86	Male	Occult	0.6	1.0
Patient 5	59	Male	Occult	0.5	1.2
Patient 6	63	Male	Occult	0.2	1.0

Predominantly: predominantly classic CNV
Occult: occult without classic CNV
D: disease eye; N: normal eye

clinical consultation, there is no significant difference in the seriousness of disease between the two types. Table 1.1 shows patients with normal eyes that have sufficient visual acuity.

This study followed the tenets of the Declaration of Helsinki regarding research involving human subjects; informed consent was obtained from all subjects. The data in this study is routinely collected at the Kitasato University Hospital when AMD is suspected. This test does not determine the risk, however.

1.3.2 Experimental Procedure

Pupil responses were measured using a PLR observation procedure and without any procedures to stimulate mydriasis, to determine the level of functionality of the melanopsin-associated photoreceptors [19, 41].

Pupil responses were recorded using Iriscorder Dual equipment (Hamamatsu Photonics, Japan) at a sampling rate of 30 Hz. This equipment is designed for observing PLRs with ipRGCs activation in accordance with the measurements taken in previous studies [17, 19], and consists of a measurement controller and a pair of goggles which have an infrared camera and an LED light source for each eye of the goggles. Therefore, the stimulus light presentation was binocular, and the recordings were also binocular. In the experiment, a long-wavelength (635 ± 5 nm) red light and a short-wavelength (470 ± 7 nm) blue light were used at two different light intensities (10 cd/m^2 and 100 cd/m^2). The light stimulation conditions were adjusted in accordance with the method of Kawasaki and Kardon [17, 19], while the measurement validity was also confirmed [13, 29]. PLRs for light stimuli were simply recorded as single trial measure. The light pulses were projected within the housing of a pair of goggles. Subjects were asked to not blink for 20 sec, while their pupil diameter was recorded. The observation period consisted of a 10 second light pulse which caused a restriction of the pupil size, followed by 10 seconds without a light pulse during which restoration of the pupil size was allowed to occur. These

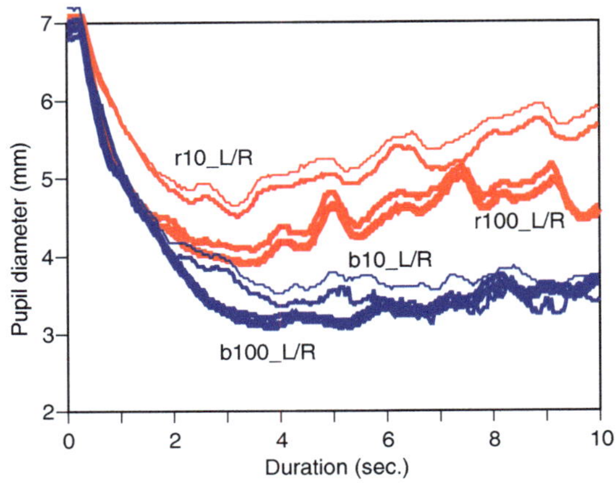

Fig. 1.1 Healthy subject reactions for both eyes

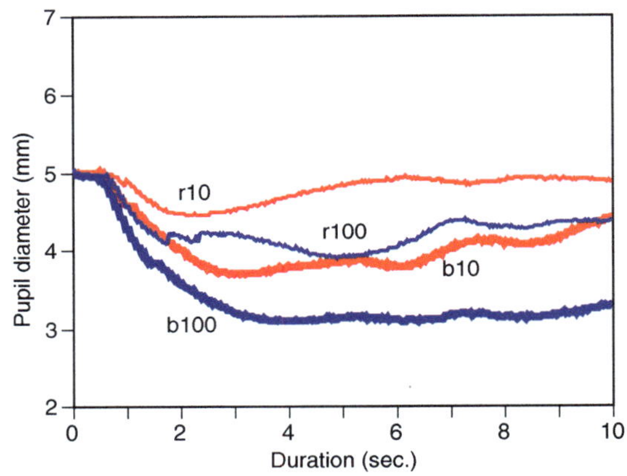

Fig. 1.2 Pupil light reflex on diseased eye of an AMD patient

measurements were taken in a dark room with constant lighting conditions. A dark adaptation period of 5 minutes was allowed prior to the taking of measurements.

Responses for both left and right pupils were recorded as pupillograms for each subject. In this work, the four conditions for the left (L) and right (R) eyes are defined as follows: "r10" (long wavelength and low light intensity), "r100" (long and high intensity), "b10" (short wavelength and low intensity) and "b100" (short wavelength and high intensity). The pupil responses of a healthy subject are illustrated in Fig. 1.1, and the pupil responses of a patient diseased eye with AMD are illustrated in Fig. 1.2. In both figures, the red line indicates pupil responses for red light, and the blue line indicates responses for blue light. In Fig. 1.1, the very fine line shows the pupil response of the left eye for 10 cd/m^2, the less fine line shows the pupil response of the right eye for 10 cd/m^2, while the bold line shows the pupil response of left eye for 100 cd/m^2 and the bolder line shows pupil response for the right eye for 100 cd/m^2.

Fig. 1.3 Averaged pupil light reflex on healthy eyes

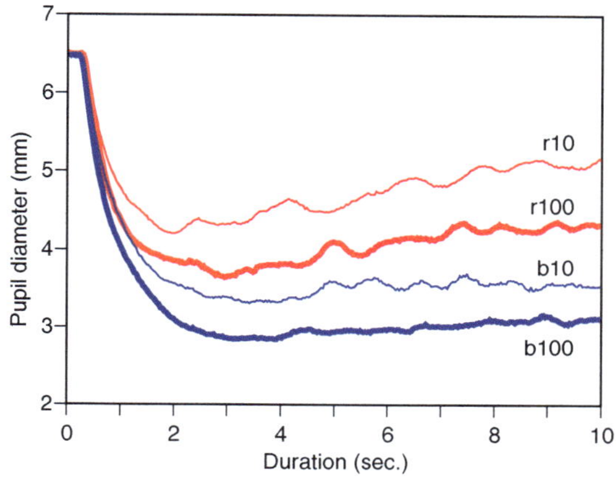

Fig. 1.4 Averaged pupil light reflex on diseased eyes

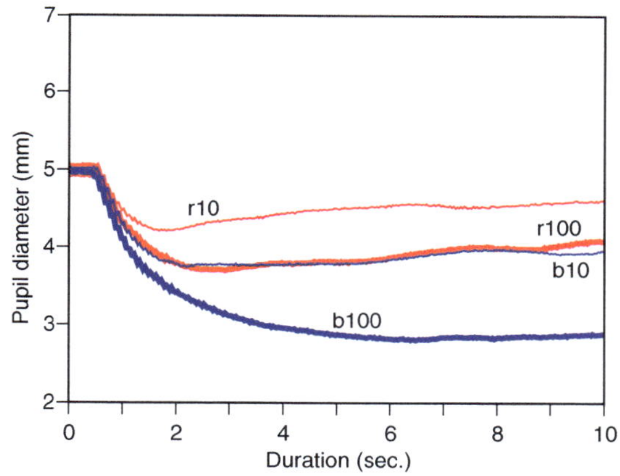

In Fig. 1.1, there is little difference in PLR waveform shapes between the left and right eyes. Pupil sizes are sustained during both blue light pulses, and restoration of pupil sizes can be observed after both red light pulses. These phenomena confirm pupil behaviour using blue and red light pulses observed in previous studies. However, these phenomena influenced diseased eyes, as shown in Fig. 1.2.

To emphasise the difference in PLR waveforms between patients' diseased eyes and healthy eyes, averaged PLRs are illustrated for all healthy eyes in Fig. 1.3 and for all diseased eyes in Fig. 1.4. Regarding Fig. 1.3, the order of mean pupil diameters clearly shows the degree of pupil restrictions for stimuli. The order for patient with diseased eyes may be influenced by the disease. Table 1.2 shows mean pupil diameters and STDs for four stimuli across healthy subjects, patients with normal eyes and patients with diseased eyes. The order of means is maintained across three types of eyes, and it does not seem easy to classify eye conditions using these statistics.

Table 1.2 Mean pupil diameters for the stimuli

	Healthy subject		Patient normal		Patient diseased	
	Mean	STD	Mean	STD	Mean	STD
r10	4.82	0.86	3.98	0.68	4.50	0.74
r100	4.30	0.90	3.41	0.73	3.99	0.74
b10	3.74	0.85	3.23	0.61	3.96	0.78
b100	3.16	0.85	2.65	0.69	3.18	0.74

During all observations for pupil reactions, repeated measures were not taken and so the test-retest reliability of the paradigm has not been evaluated.

1.3.3 Fourier Descriptors

In order to extract measurements of formed objects, conventional Fourier descriptors are introduced. This technique can be applied to objects of various shapes. The issue in this work is characterising waveform shapes, and this technique can be applied, as Fourier Transform is often used to analyse signal waveforms. The feature vectors for PLR waveforms were extracted using the Discrete Fourier Transform (DFT) procedure [32, 43]. As mentioned above, PLRs were sampled as discrete signals. Here, the length N of a discrete signal is defined as $x(n)$, which is sampled at time t with spacing Δ. The signal $x(n)$ can be noted as an Eq. (1.1) using DFT [26].

$$x(n) = a_0 + \sum_{k=1}^{N/2} \left(a(k)cos\left(2\pi k \frac{t(n)}{N\Delta}\right) + b(k)sin\left(2\pi k \frac{t(n)}{N\Delta}\right)\right) \tag{1.1}$$

$$a_0 = X(1)/N$$

$$a(k) = 2\,real(X(k+1))/N$$

$$b(k) = 2\,imag(X(k+1))/N$$

This suggests that PLR waveforms can be represented using coefficients a_0, $a(k)$ and $b(k)$ with periodical sine and cosine functions. To present the features of the waveforms of pupillograms, the magnitudes of the coefficients are preferred, because coefficient $b(k)$ is the imaginary part of a value. The magnitudes of coefficients, including a_0, $FD_i (i = 0, \ldots, N/2-1)$, are used as Fourier descriptors (FDs) in vector (2.2) as follows:

$$f = [FD_0, FD_1, \ldots, FD_{N/2-1}] \tag{1.2}$$

In general, the components FD_0 and a_0 in Eq. (1.1) show the DC components of the signal. These DC components represent the amplitude; however the waveform shape consists of frequency components. This means that the DC components indicate pupil diameters which are dependent on individual pupil sizes of overall average. Also, the AC components at lower frequencies are influenced by DC components. Therefore, the features are affected by individual factors, so that a standardised feature using a component, such as FD_1 for example, is preferred in vector (1.3), which is converted from the above vector (1.2), as follows [43]:

$$f = \left[\frac{FD_2}{FD_1}, \frac{FD_3}{FD_1}, \ldots, \frac{FD_{N/2-1}}{FD_1}\right] \tag{1.3}$$

The vector represents a basic feature of waveform shapes of PLRs. Using these types of vectors for measurements, the responses given under experimental conditions can be compared.

1.3.4 Prediction Procedure for AMD

This work proposes a procedure for detecting AMD patients and diseased eyes. The procedure is summarised in a flowchart in Fig. 1.5. The first step is the definition of feature vectors of PLR waveforms for four stimuli using the FD technique mentioned above. Both eyes should be observed. The second step is to measure the (dis)similarity as distances between feature vectors of PLR waveforms for individuals for each eye. Finally, prediction in this work is conducted using a multi-dimensional scaling (MDS) technique to characterise vector differences, including differences between individuals, and using a random forest (Random Forest) technique as a classification procedure. The process of analysis is represented as follows.

Processing and prediction can be conducted for a few subjects without repeated measurements. Of course, the validity may not be sufficient, but it can be improved

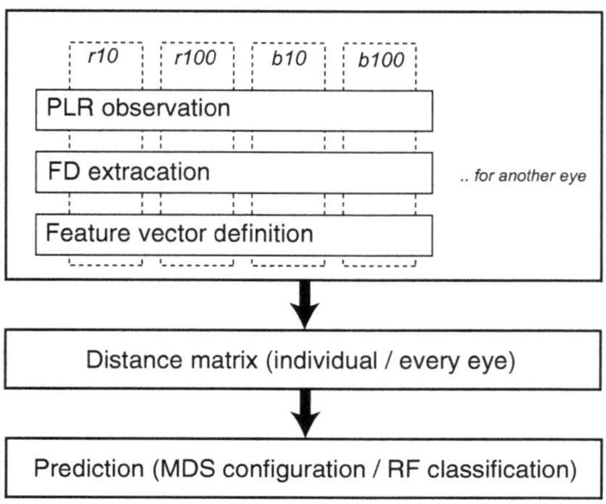

Fig. 1.5 Flowchart of signal processing and prediction

step by step when data from additional subjects is gathered. In this way, all of the techniques employed in this procedure are simple. This work is a feasibility study that examines the possibility of screening patients. A more suitable procedure will be developed once this possibility has been confirmed.

1.4 Feature Descriptions

1.4.1 Feature Definition

The features of observed PLRs are presented using the procedure specified in the Experimental Procedure section. The actual calculations were conducted using MATLAB (Mathworks, Inc.). To extract features of pupil constriction in transient and sustained pupil reactions, pupil responses mainly during the first 10 seconds are analysed because they are melanopsin-based reactions. Every individual observation of the pupillogram for a stimulus was set to a signal $x(n)$ in Eq. (1.1); then FD vectors were extracted such as in Eq. (1.2).

First, FD_0 were extracted in order to compare waveform amplitudes which were extracted from the transform. The amplitudes for PLR waveforms of all responses were calculated. Here, the amplitudes were standardised using the maximum peaks of the waveforms, such as b100, which suppresses individual differences for example [17]. The means were summarised into three categories: patients with diseased eyes, patients with normal eyes and healthy eyes. The results are shown in Fig. 1.6. The error bars indicate standard errors. According to previous studies and Fig. 1.1, the amplitude increases from the "r10" condition to the "b100" condition. The mean amplitude of healthy eyes responds to the order. However, these orders are influenced in patients with one normal and one diseased eye. Regarding Figs. 1.1 and 1.2, pupil diameters at stimulus onset as baselines are different. The mean diameter for patients is 4.87 mm, and the mean for healthy subjects is 6.46 mm,

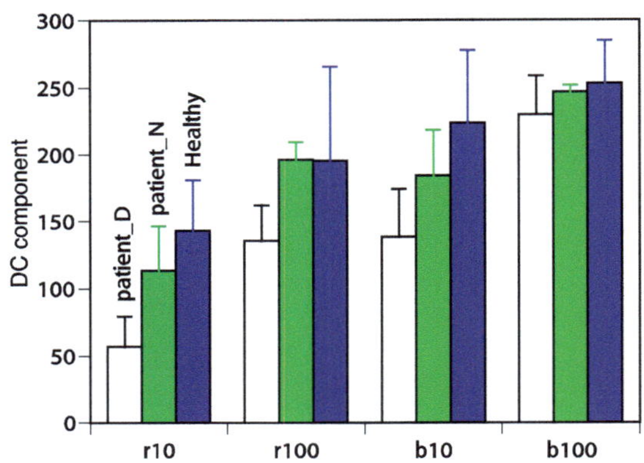

Fig. 1.6 Amplitudes of waveforms

while there is a significant difference between them ($t(10) = 4.88, p < 0.05$). The factor of this difference is not determined, however. There is no significant difference in mean pupil diameters between patients with diseased and normal eyes ($t(10) = 0.60, p = 0.56$). The above result shows that the amplitudes present differences which are independent of baseline pupil diameters, since the differences change with the stimulus conditions in Fig. 1.6.

The waveform shape can be noted as a vector in the Fourier Descriptors section. A number of components of the feature vector represent the characteristics of most signals at only the low-order values of 4 or 5 FDs [32]. Therefore, four dimensions have been employed as feature vectors (f) in this book. Also, FD_0 was too large in this observation, so FD_1 was used for the standardisation. As an example, Fourier descriptors of "r10" for the left eye of the healthy subject in Fig. 1.1 are given as Eq. (1.2):

$$f_{r10\_L} = [501.0, 83.6, 50.6, 28.2, 23.3, 30.0, \dots] \tag{1.4}$$

This vector is converted using Eq. (1.3):

$$f_{r10\_L} = \left[\frac{50.6}{83.6}, \frac{28.2}{83.6}, \frac{23.3}{83.6}, \frac{30.0}{83.6}\right] \tag{1.5}$$

Therefore, the feature vectors are noted as follows:

$$\begin{aligned} f_{r10\_L} &= [0.61, 0.34, 0.28, 0.36] \\ f_{r100\_L} &= [0.68, 0.52, 0.40, 0.36] \\ f_{b10\_L} &= [0.68, 0.47, 0.48, 0.33] \\ f_{b100\_L} &= [0.64, 0.43, 0.33, 0.30] \end{aligned} \tag{1.6}$$

1.4.2 Similarity/Dissimilarity

As both Figs. 1.1 and 1.2 show, the waveforms of PLRs in response to stimuli are different between subjects. To compare the shapes of waveforms quantitatively, the metrics of similarity and dissimilarity should be defined using distance metrics and waveform feature vector which are noted above. This is a very popular approach for pattern recognition, such as categorisation and discrimination of waveforms [37]. Here, the Euclidean distance (or Minkowski's power metric) can be defined as the Euclidean norm between two feature vectors. This is the dissimilarity metric of distance measurements of feature vectors of PLR waveform shapes. This Euclidean distance is simply called distance, as follows. The distances between stimuli conditions mentioned above for a set of PLRs of an eye (left eye) are summarised in the following matrix (Ed_L) as a triangular matrix.

$$Ed\_L = \left(\begin{array}{c|cccc} & r10 & r100 & b10 & b100 \\ \hline r10 & 0 & 0.23 & 0.25 & 0.12 \\ r100 & 0.23 & 0 & 0.10 & 0.13 \\ b10 & 0.25 & 0.10 & 0 & 0.16 \\ b100 & 0.12 & 0.13 & 0.16 & 0 \end{array} \right) \tag{1.7}$$

These components indicate dissimilarity between the two waveform shapes when their amplitude factors are excluded.

1.5 Comparing PLRs Between Eyes

1.5.1 Features of Two Eyes

There are many cases where one of the eyes is diseased and the other is not. Table 1.1 shows visual acuities of both diseased(D) and normal(N) eyes of patient participants. All patients have a level of visual acuity in one eye that is comparable to healthy people, and a diseased eye with poor acuity. Regarding the progress of AMD, the disease seldom progresses simultaneously in both eyes. This means that there is a difference in retinal condition between the left and right eyes of most patients with AMD. Since the retinal condition is different, the PLR waveforms are also different, as shown in Figs. 1.1 and 1.2, while both healthy eyes respond similarly in Fig. 1.1. This phenomenon suggests that quantitative differences in PLRs between two eyes may provide some information regarding symptoms of AMD. Therefore, a procedure of feature extraction from PLRs and the creation of a distance matrix in the above section can be applied to other cases with two eyes.

Here is a set of feature vectors of PLR waveform shapes of a patient:

$$\begin{aligned} f_{r10\_D} &= [0.54, 0.22, 0.21, 0.17] \\ f_{r100\_D} &= [0.62, 0.38, 0.30, 0.24] \\ f_{b10\_D} &= [0.59, 0.37, 0.25, 0.20] \\ f_{b100\_D} &= [0.69, 0.49, 0.37, 0.30] \\ f_{r10\_N} &= [0.65, 0.23, 0.23, 0.14] \\ f_{r100\_N} &= [0.55, 0.42, 0.19, 0.08] \\ f_{b10\_N} &= [0.23, 0.56, 0.29, 0.26] \\ f_{b100\_N} &= [0.72, 0.48, 0.36, 0.28] \end{aligned} \tag{1.8}$$

In this notation, D means diseased eye, and N means normal eye for each experimental stimulus condition, such as "r10".

Regarding the procedure and feature vectors, a distance matrix across two eyes for one subject can be created as follows:

$$
Ed\_D/N = \left(\begin{array}{c|cccccccc}
 & r10\_D & r100\_D & b10\_D & b100\_D & r10\_N & r100\_N & b10\_N & b100\_N \\ \hline
r10\_D & 0 & & & & & & & \\
r100\_D & 0.23 & 0 & & & & & & \\
b10\_D & 0.55 & 0.40 & 0 & & & & & \\
b100\_D & 0.32 & 0.32 & 0.50 & 0 & & & & \\
r10\_N & 0.11 & 0.22 & 0.47 & 0.36 & 0 & & & \\
r100\_N & 0.19 & 0.21 & 0.43 & 0.16 & 0.21 & 0 & & \\
b10\_N & 0.17 & 0.15 & 0.41 & 0.22 & 0.17 & 0.07 & 0 & \\
b100\_N & 0.34 & 0.32 & 0.47 & 0.04 & 0.36 & 0.16 & 0.22 & 0
\end{array}\right) \tag{1.9}
$$

To maintain the presentation of distance matrix for two eyes, both the left and right eyes (L/R) are allotted instead of diseased and normal eyes (D/N) for healthy subjects since they have two healthy eyes.

The distances between PLR waveforms for subjects with normal eyes are relatively shorter than the distances between diseased eyes and distances between one normal and one diseased eye. A matrix can be created for every subject based on this procedure.

According to the matrix of one patient, the distances representing the dissimilarity between the conditions for diseased eyes are longer than the distances for one normal eye and one diseased eye and the distances for two normal eyes. The distances for subjects with two healthy normal eyes are the shortest. Distance matrices were created for all subjects.

1.5.2 Configurations Using MDS

To create an overall structure of the relationship between PLR waveforms, the Multi-Dimensional Scaling (MDS) method [38] was applied to the distance matrix [27].

The basic approach of MDS is as follows:

There are n samples that have feature vectors of waveforms, and the distance can be defined using the Euclidean distance such as o_{ij} between samples i and j. The diagonal distance components are 0.

Here, an n by A matrix X of the waveform coordinates is introduced, and another distance d_{ij} can be defined as the following equation:

$$d_{ij} = \left\{ \sum_{a=1}^{A} (x_{ia} - x_{ja})^2 \right\}^{1/2} \tag{1.10}$$

Additionally, the monotonic transformation of the data g is introduced, and the following equation is minimised:

$$\phi(g, \mathbf{X}) = \sum_{j<i} (g(o_{ij}) - d_{ij})^2 \tag{1.11}$$

As a result, MDS produces a low-dimensional projection of the data which can present the paired distances between data points. As the low-dimensional projection provides a way to configure the data, the illustration can be used to visually understand the relationship between the data. A simple MDS technique is based on measurable distance scales such as space of Euclidean. A well-known example is the reproduction of a geographical map from distance information between cities. Using a matrix of distances between cities, the MDS technique can find appropriate geographic locations. Simple comparison of distances between cities may not adequately represent their geographical locations. The MDS technique first transforms the data into inner products which indicate cosine similarities, after which eigenvalues are calculated and vector decomposition is conducted to extract lower level dimensional information as spatial representations. The monotonic transformation is noted as g in the equation above [38]. The measurements can be extended to response frequencies. A same-different task was conducted using pairs of colour samples, and correct discrimination probabilities between colours were summarised as dissimilarity measurements. In applying the MDS technique to the data set, the extracted features of the colours can represent a relationship such as a colour wheel when used as a system of colour perception. These samples suggest that the MDS technique is not only for extracting a set of features of data but also to provide information about a latent data structure. This technique has been employed to classify individuals as patients or as in the normal control group when their responses deviate from the normal response structure.

The Individual Difference MDS procedure has been introduced to extend conventional MDS analysis to multiple distance matrices using the features of every subject [2, 20]. This procedure is known as individual difference MDS (ID MDS). The above equation is modified by introducing subject k to represent o_{ijk}, which indicates the similarity between two samples i and j from subject k. The distances are updated in the following equation:

$$d_{ijk} = \left\{ \sum_{a=1}^{A} w_{ka} (x_{ia} - x_{ja})^2 \right\}^{1/2} \tag{1.12}$$

Once again, x_{ia} is the coordinate of sample i on dimension a, and w_{ka} is the weight of the dimension a for subject k. Matrix $\mathbf{W}$ is introduced to indicate the

Table 1.3 Three dimensional information of MDS for stimuli

	D/N			H		
	dim1	dim2	dim3	dim1	dim2	dim3
r10	−0.89	0.16	0.24	−0.71	0.17	−0.22
r100	0.27	0.26	0.30	0.35	0.37	−0.09
b10	0.21	−0.38	0.33	0.18	−0.44	−0.13
b100	0.33	−0.05	−0.17	0.25	−0.08	−0.26

D/N: diseased and normal eyes of AMD patients
H: healthy eyes

individual differences in weights of the features of individual subjects [38]. As another example mentioned above, a same-different decision task for pairs of colour samples of several participants was measured twice. The ID MDS technique extracts the weights of *a* dimension for different subjects, in addition to representing the relationship between the colour samples. As noted above, individual difference information, including participant characteristics, can be extracted as comparable measurements. In this work, the characteristics of both participants and stimuli were extracted and evaluated using the ID MDS technique. The actual calculation was conducted using R. During MDS analysis, the number of dimensions was set to 3. Again, the dimensions are defined in order to present a low-dimensional projection regarding MDS calculations. Therefore, the dimension may present the features of the spatial layout of the data. The dimensional values are summarised in Table 1.3. The contributions of both Dimension 1 and Dimension 2 to the classification have the same tendency as in the case of a two-dimensional analysis. Following this Dimensions 1 and 3 were then compared. The stimulus is configured in a two-dimensional space which was created using MDS analysis, as shown in Fig. 1.7. The horizontal axis shows Dimension 1, and the vertical axis shows Dimension 3. According to the distance matrix, one of the waveforms of a stimulus includes a diseased and a normal eye (D/N), and the other includes eyes that are both healthy (H). The conditions are indicated in Fig. 1.7 as "D/N" for the former and "H" for the latter. All normal conditions are gathered in one cluster except "r10H", and only "b100-D/N" belongs to a cluster which consists of normal responses. This suggests that for normal responses stimuli conditions are configured outside of the cluster when the conditions include abnormal responses. This means that the possibility of detecting abnormal responses exists. According to the configuration, it is interesting that the r10 condition always shows a different tendency. This stimuli condition may provide significant information about the abnormal responses of diseased eyes.

All results for the eight conditions of the six healthy subjects and six patients are mapped in Fig. 1.8. The stimuli conditions produce clusters in response to the configurations of stimuli, as shown in Fig. 1.8 where all subjects' data is mapped in a similar style. However, the plots for the patient subjects are positioned in a different area. In particular, the values for patient subjects deviate from the norm in Dimension 3. Even the plots of normal eyes of patient subjects are more widely distributed. The features of most patients show a different tendency. All subjects can be configured using their own individual two-dimensional information, as shown in

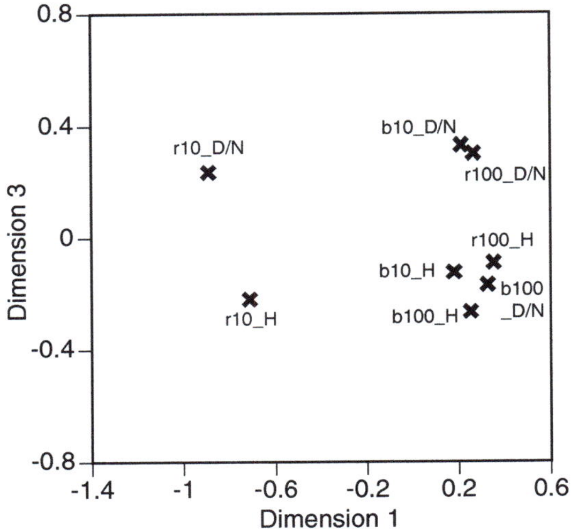

Fig. 1.7 Light pulse condition configurations using three dimensional scales

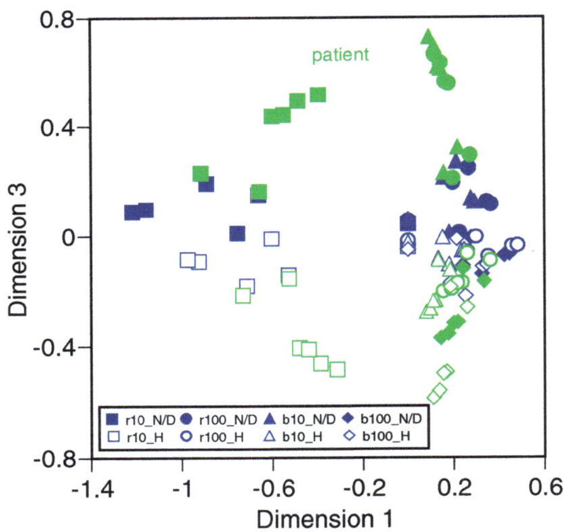

Fig. 1.8 The conditions for subjects configured using three dimensional scales

Fig. 1.9. In this figure, the value of Dimension 3 clearly indicates the differences between healthy subjects and patient subjects. The healthy subjects produce a cluster in the lower area, and the patient subjects' plots are distributed around the cluster. If the borders between the classes could be defined mathematically, classification according to health condition may also be possible. The meaning of Dimension 3 can not be defined mathematically in response to the nature of MDS analysis, however. A detailed analysis of the representation in this topic will be a subject of our further study.

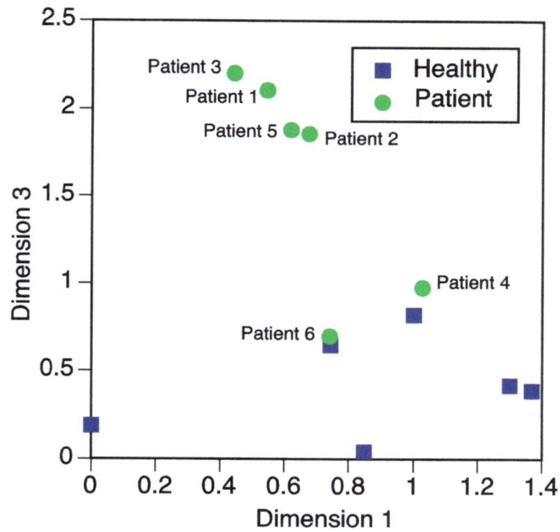

Fig. 1.9 The subjects' configurations using three-dimensional scales

Since MDS analysis also provides a three-dimensional feature for every subject and patient, grouping of participants is possible using cluster analysis and data of the feature set. The extracted three-dimensional features may represent individual attribution and can be used to classify participants. Cluster analysis is introduced in order to classify feature vectors. The dendrogram as a result of cluster analysis using the Ward method is summarised in Fig. 1.10. The horizontal axis shows averaged distance between subjects. The clustering process responds to the distribution of subjects in Fig. 1.7. Most patients (the predominant type) are classified apart from other subjects.

This procedure can illustrate some of the differences between healthy subjects and patient subjects, and the clustering results show the tendency towards deviations. Therefore the subject categories (healthy or patient) are ambiguous. Regarding the specific purpose of this work, more quantitative classification may be required as a result.

In the next section, a possible procedure for this will be introduced and discussed.

1.6 Classification Using a Combination of Amplitudes and Features of Waveform Shapes of Pupillograms

1.6.1 Feature Combinations

In the previous section, the features of waveform shapes were extracted as amplitudes and Fourier descriptors. There were significant differences in the mean amplitudes between the eye conditions when the features of waveform shapes were

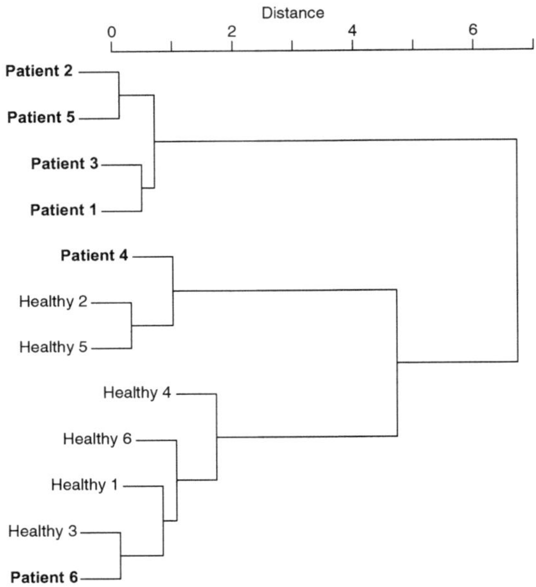

Fig. 1.10 Results of clustering of all subjects

used to separate the subjects. Though they can indicate some of the conditions affecting eyes, they are not used collaboratively. Since they are recognised as different scales of features, some consideration is required in order to combine them. Also, a more robust classification technique should be considered. As noted above, the features of waveform shapes are based on frequency components of pupillary changes without the DC component of pupil size. The initial size of the pupil represents an ordinal condition, and it also affects the range of change of PLR. Therefore, the initial pupil size, represented as FD_0, should be considered as a feature.

To include FD_0 components with features, the FD_0 values were standardised using the means for each eye. In the case of the left eye of the healthy subject in Fig. 1.1, the $FD_0{}'$ of the standardised FD_0 uses a mean of components. Both are noted as follows:

$$FD_{0:L} = [FD_{0:r10\_L}{}' FD_{0:r100\_L}{}' FD_{0:b10\_L}{}' FD_{0:b100\_L}{}'] \tag{1.13}$$

$$[FD_{0:r10\_L}{}' FD_{0:r100\_L}{}' FD_{0:b10\_L}{}' FD_{0:b100\_L}{}'] = [0.65, 0.95, 1.16, 1.25]$$

$$FD_{0:L}{}' = \frac{FD_{0:L}}{\frac{1}{n}|FD_{0:L}|} \tag{1.14}$$

$$FD_{0:L}{}' = [0.65, 0.95, 1.16, 1.25]$$

Combined vector f' can be noted as a modified feature vector using weight coefficient w.

$$f' = \left[\frac{FD_2}{FD_1}, \frac{FD_3}{FD_1}, \ldots, \frac{FD_5}{FD_1}, wFD_0{}'\right] \tag{1.15}$$

Here, w is a coefficient used as a weighted value to create a balance between the features of waveform shapes and standardised FD_0 ($FD_0{}'$). To optimise the coefficient w, the performance of this classification is evaluated in response to variations of w in the section which follow.

1.6.2 *Classification Procedure*

In this work, the number of subjects is small, and the number of trials for taking measurements is limited because the experimental stimuli influences the response. The Random Forest method [24], which uses an ensemble learning procedure to analyse the classification of a small sample, is used frequently. This technique is a kind of machine learning and data mining procedure, which is based on a decision tree for classification. As the generalisation error is estimated during analysis, an optimisation can be conducted easily. A tuned model can be used to make predictions with other sets of data. Determination of the feasibility of the classification model may be useful in our future work. Also, the Random Forest method can show contributions of features and can be used to conduct cross-validation calculations. Since the structure of the feature data set is not clear, the results of the Random Forest method may provide useful information to improve the detection procedure. The statistical package R and the "RandomForest" package [1, 23] were used for this analysis.

The number of decision trees was set at 500 as a default value, and the sample size was set at 6. The total number of samples was 24 ($2 \times (6 + 6\ subjects)$). One-third of the samples were assigned as test data, and the rest of data was assigned as OOB (Out of Bag) training data. The selection of the data set was initially random. This selection was performed ten times, in order to calculate the generalised performance of the data throughout all conditions. To optimise the coefficient w, performance was evaluated according to the value of w, as follows:

$$w = [0, 0.3, 0.5, 1.0, 1.5, 2.0, 3.0, \infty] \tag{1.16}$$

Here, ∞ means a case using only $FD_0{}'$.

Classification of the modified set of features into two classes, healthy and normal (HN) and AMD (D), or into three classes, healthy (H), AMD patient (D) and patient normal age-affected eyes (N), was conducted using the RF technique.

According to the preliminary analysis, the performance was low when the feature set of equation (1) was used. Next, Euclidean distances between the four conditions were analysed in the same way as in the previous assessment [28].

Here, a Euclidean distance matrix is shown as $Ed_{Healthy\ L}$ for the example of the healthy subject's left eye, and the distance feature is noted as FE. The FE vector consists of distance components without zero distance such as diagonal components in Ed matrix.

$$Ed_{Healthy\ L} = \left(\begin{array}{c|cccc} & r10 & r100 & b10 & b100 \\ \hline r10 & 0 & 0.37 & 0.57 & 0.62 \\ r100 & 0.37 & 0 & 0.24 & 0.33 \\ b10 & 0.57 & 0.24 & 0 & 0.18 \\ b100 & 0.62 & 0.33 & 0.18 & 0 \end{array}\right) \quad (1.17)$$

$$FE = [r10 - r100, r10 - b10, r10 - b100, r100 - b10, r100 - b100, b10 - b100] \quad (1.18)$$

$$FE_{Healthy\ L} = [0.37, 0.57, 0.62, 0.24, 0.33, 0.18]$$

1.6.3 Two-Class Performance

For two-class classification of PLRs of healthy and normal (HN) and AMD-affected diseased (D) eyes, ten times calculations were conducted using w values. Mean error rates of classifications are summarised in Fig. 1.11. The horizontal axis shows weight w, and the vertical axis shows the error rate. The error bar in the figure shows the standard deviation (STD). The minimum error rate appears at $w = 1.5$. In the case of $w = 0.5$ or only FD_0', the error rates are high.

The estimation performance for $w = 1.5$ is summarised as a contingency table in Table 1.4. The horizontal cells (columns) show the prediction categories, and the vertical cells (rows) show the categories of the participants. Diagonal components show correct classifications. The table shows that healthy and normal (HN) eyes can almost always be correctly classified, and though the error rate of classification for AMD-affected diseased eyes (D) is low, at 17%, overall performance seems not bad for the cases observed.

1.6.4 Three-Class Performance

For three-class classification of PLRs of healthy (H), AMD-affected disease (D) and age-affected normal (N) eyes, ten times calculations were conducted using w values. Mean error rates across w values are summarised in Fig. 1.12 using the same

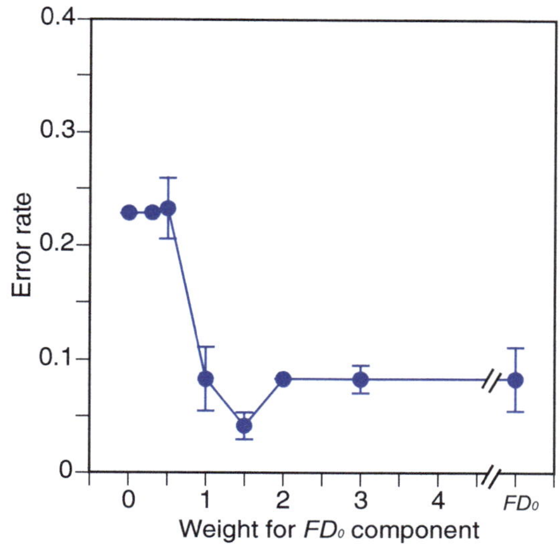

Fig. 1.11 Two-class classification error rate across w values

Table 1.4 Contingency table for two classes

	D	HN	Err.
AMD (D:n=6)	5	1	0.17
Healthy and normal (HN:n=18)	0	18	0.00

(w for $FD_0{}' = 1.5$)

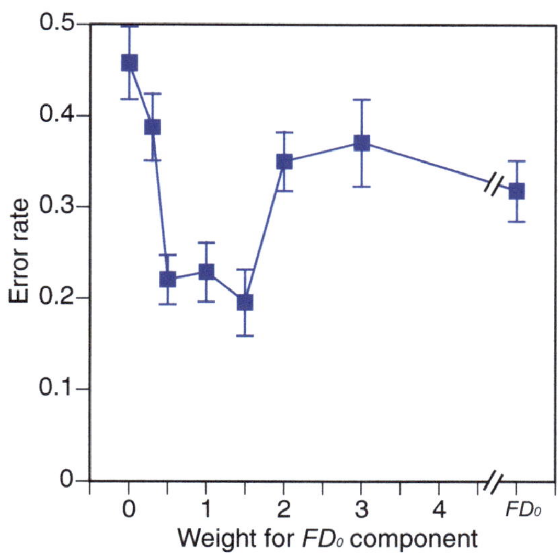

Fig. 1.12 Three-class classification error rate across w values

format as in Fig. 1.11. Error bars in the figure show the STDs of all ten results. The minimum error rate also appears when $w = 1.5$, while the rate changes with the w values.

Table 1.5 Contingency table for three classes

	D	N	H	Err.
AMD (D:n=6)	5	0	1	0.17
Normal (N:n=6)	1	3	2	0.50
Healthy (H:n=12)	0	0.7	11.3	0.06

(w for $FD_0{}' = 1.5$)

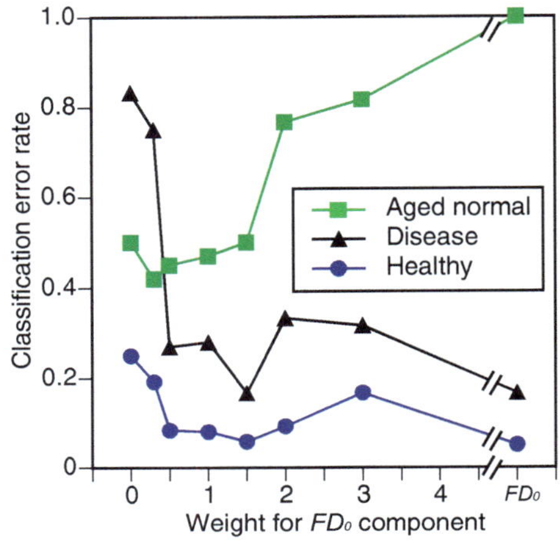

Fig. 1.13 Error rates for classification of each class

The results for three classifications are summarised as a contingency table in Table 1.5, as mean rates for all ten results. According to the results, healthy eyes are almost always correctly classified, while the error rate for AMD-affected diseased eyes is once again 17%. The performance for age-affected normal eyes is not good, as the error rate is over 50%. The age-affected normal eye class may include eyes which respond in the same manner as healthy eyes or eyes which are affected by AMD. While the extracted features may represent a hybrid combination of both categories of eyes, classification performance remained at the chance level as a result. The feature characteristics of every patient should be examined. Therefore, further observation of the condition of each subject's eye may be required.

The dependency of w for three-class classifications is examined using a simple comparison of classification errors of the three categories. The error rates for the three-class classifications are summarised in Fig. 1.13. The rates of both healthy and AMD-affected eyes are almost always small, and the minimum rate appears at $w = 1.5$. The rates for age-affected eyes are relatively higher than the ones for the other two classes. The rates for age-affected eyes increase with w, and the features of PLR waveform shapes may contribute more to classification than the factor F_0. More detailed appropriate features may help to improve performance.

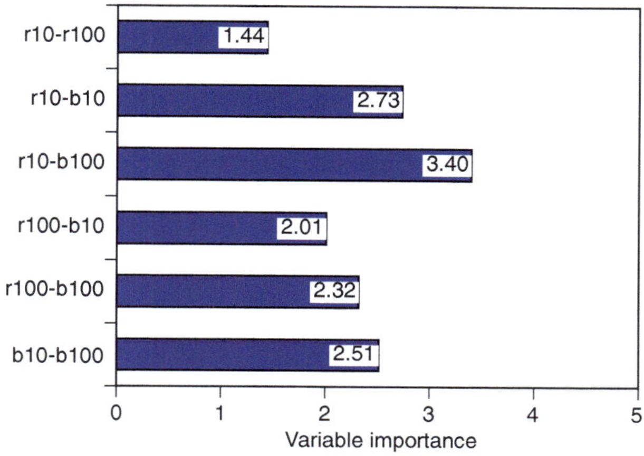

Fig. 1.14 Comparison of contribution values across features (w=1.5)

1.6.5 Features Contributing to Estimations

According to the classification results in the previous section, pupils can be classified accurately using features of PLR waveforms, except for the performance of age-affected normal eyes.

The next question was which components of features make it possible to classify PLRs. After that, the contributions of features were evaluated using the Random Forest tool. The degree of contribution of each feature or the Gini coefficients of the features are summarised in Fig. 1.14 as an index of their contribution. The figure suggests that Euclidean distances are dissimilar between PLRs for blue light and PLRs for red light at low intensities such as $r10 - b100$ or $r10 - b10$ and denote the level of performance. Also, colour differences at high intensities in PLRs such as $r100 - b100$ or $b10 - b100$ present the ability to distinguish the condition of the pupil.

These results coincide with the visual differences in PLRs between healthy and AMD-affected patients. More detailed features of these differences should be extracted carefully in order to facilitate more precise classification.

1.6.6 Clustered PLRs

The similarities of the features of waveform shapes in PLRs were also identified using the Random Forest procedure, and cluster analysis of these distances was then conducted using the Ward method. Figure 1.15 shows the resulting dendrogram of the clusters of PLR features with $w = 1.5$ as the optimised value. In this figure, all subjects and eyes are indicated as healthy/patient and left/right or normal/diseased.

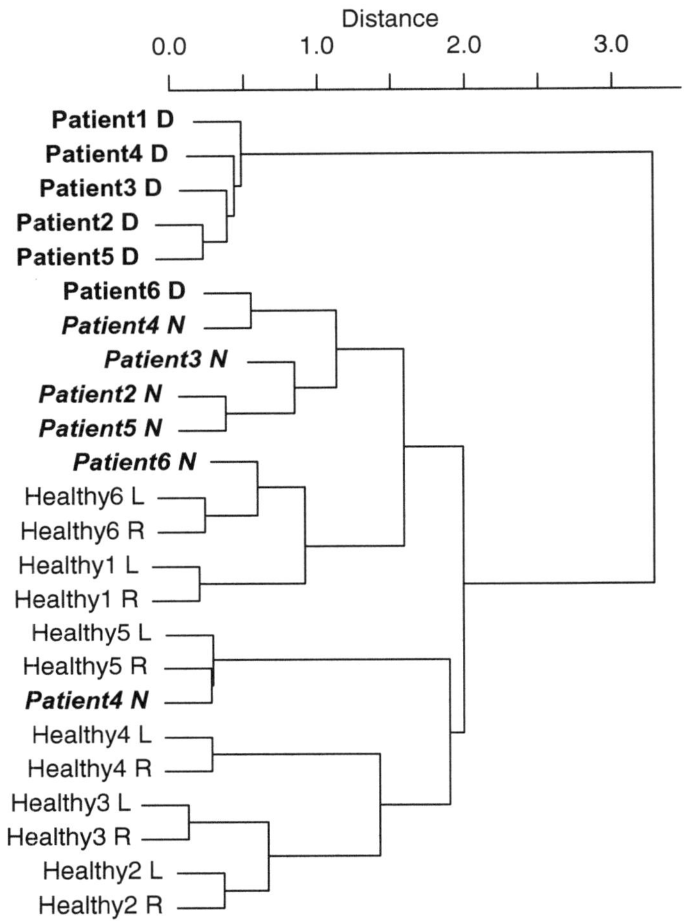

Fig. 1.15 Results of clustering of PLRs of all subjects (w=1.5)

As the figure shows, the lower cluster displays a group of healthy subjects except for two patients with age-affected eyes. A sub-cluster consists of both eyes of most subjects. The upper clusters consist of patients with age-affected eyes in the lower part and patients with AMD-affected eyes in the upper part. Some healthy subjects and some patients with AMD-affected eyes have been incorrectly classified into these groups. These occurrences have been explained in the above section where performance was classified and evaluated.

To improve the performance of the test, responses should be ophthalmologically diagnosed. For the classification procedure, many different data mining tools have been developed, so a more effective procedure should be devised. The characteristics of features of pupillary response waveform shapes of pupillograms which result from using the Random Forest method will be a subject of our further study.

1.7 Conclusions

This chapter proposes a possible procedure for detecting AMD-affected eyes and age-affected eyes using features of PLR waveforms. The Fourier descriptor technique was applied to extract the features of PLR waveform shapes of pupillograms and their amplitudes.

To detect affected patients using the extracted features, Multi-Dimensional Scaling (MDS) and clustering techniques were applied to emphasise certain stimuli and subject differences, since there were restrictions on gathering patient data. The performance is shown as a qualitative tendency.

To quantitatively evaluate detection performance, an appropriate classifier such as the Random Forest method was introduced. For this classification, a combination of features was used in order to obtain detailed features of waveform shapes of pupillograms. This was a balanced combination of the two features and was controlled using weighted values. As a result of classification analysis using the Random Forest method, the performance of the three categories (patients, aged and healthy subjects) was compared after the weighted values were optimised using variations of the classification error rate.

The results show that the error rates for healthy pupils and AMD-affected pupils were low when the value of the coefficient for a combination of PLR amplitudes and feature of waveforms was optimised as 1.5. However, the error rates for patients with age-affected eyes were not low.

The pupils of patients with age-affected eyes were influenced by various factors, so it may not be easy to classify healthy and AMD-affected patients in some cases. Additional feature processing, such as processing of the pupil diameter [33, 34] and having a sufficient number of patients for statistical analysis [7], may be required.

Also, the evaluation of additional diagnostic procedures will be a subject of our further study in the future. In particular, feature extraction issues for pupillary responses are often suggested. As noted in the related works, pupil responses have been discussed as an index for some patients or conditions. More appropriate evaluation methodology may provide an accurate diagnostic procedure. An overall technique for measurement of pupil response should be considered as a subject of our further study.

References

1. Package: randomForest. http://stat-www.berkeley.edu/users/breiman/RandomForests
2. Package: smacof. http://cran.r-project.org/web/packages/smacof/smacof.pdf
3. Beatty, J.: Task-evoked pupillary response, processing load, and the structure of processing resources. Psychol. Bull. **91**(2), 276–292 (1982)
4. Brozou, C.G., Fotiou, D., Androudi, S., Theodoridou, E., Giantselidis, C., Alexandridis, A., Brazitikos, P.: Pupillometric characteristics in patients with choroidal neovascularisation due to age-related macular degeneration. Eur. J. Ophthalmol. **19**(2), 254–262 (2009)

5. Dacey, D.M., Liao, H.W., Peterson, B.B., Robinson, F.R., Smith, V.C., Pokorny, J., Yau, K.W., Gamlin, P.D.: Melanopsin-expressing ganglion cells in primate retina signal color and irradiance and project to the LGN. Nature **433**, 749–754 (2005)
6. Feigl, B.: Age-related maculopathy—linking aetiology and pathophysiological changes to the ischaemia hypothesis. Prog. Retin. Eye Res. **28**, 63–86 (2009)
7. Feigl, B., Mattes, D., Thomas, R., Zele, A.J.: Intrinsically photosensitive (melanopsin) retinal ganglion cell function in glaucoma. Invest. Ophthalmol. Vis. Sci. **52**(7), 4362–4367 (2011)
8. Feigl, B., Zele, A.J.: Melanopsin-expressing intrinsically photosensitive retinal ganglion cells in retinal disease. Ophthalmol. Vis. Sci. **91**(8), 894–903 (2014)
9. Gamlin, P.D., McDougal, D.H., Pokorny, J.: Human and macaque pupil responses driven by melanopsin-containing retinal ganglion cells. Vis. Res. **47**, 946–954 (2007)
10. Hatori, M., Panda, S.: The emerging roles of melanopsin in behavioral adaptation to light. Trends Mol. Med. **16**(10), 435–446 (2010)
11. Hattar, S., Liao, H.W., Takao, M.: Melanopsin-containing retinal ganglion cells: architecture, projections, and intrinsic photosensitivity. Science **295**, 1065–1070 (2002)
12. Hess, E.H., Polt, J.M.: Pupil size in relation to mental activity during simple problem solving. Science **143**(3611), 1190–1192 (1964)
13. Ishikawa, H., Onodera, A., Asakawa, K., Nakadomari, S., Shimizu, K.: Effects of selective-wavelength block filters on pupillary light reflex under red and blue light stimuli. Jpn. J. Ophthalmol. **56**, 181–186 (2012)
14. Ishikawa, S., Ukai, K.: Pupillary escape and visual fatigue phenomena in optic nerve disease. Adv. Diagnostic Visual Optics **41**, 187–195 (1983)
15. Kankipati, L., Girkin, C.A., Gamlin, P.D.: Post-illumination pupil response in subjects without ocular disease. Visual Neurophys. **51**(5), 2764–2769 (2010)
16. Kankipati, L., Girkin, C.A., Gamlin, P.D.: The post-illumination pupil response is reduced in Glaucoma patients. Visual Neurophys. **52**(5), 2287–2292 (2011)
17. Kardon, R.H., Anderson, S.C., Damarjian, T.G., Grace, E.M., Stone, E., Kawasaki, A.: Chromatic pupil responses -preferential activation of the melanopsin-mediated versus outer photoreceptor-mediated pupil light reflex. Ophthalmology **116**(8), 1564–1573 (2009)
18. Kardon, R.H., Anderson, S.C., Damarjian, T.G., Grace, E.M., Stone, E., Kawasaki, A.: Chromatic pupillometry in patients with retinitis pigmentosa. Ophthalmology **118**(2), 376–381 (2011)
19. Kawasaki, A., Kardon, R.H.: Intrinsically photosensitive retinal ganglion cells. J. Neuroophthalmol. **27**, 195–204 (2007)
20. Leeuw, J., Mair, P.: Multidimensional scaling using majorization: Smacof in r. J. Stat. Softw. **31**(3), 1–30 (2009)
21. Lüdtke, H., Wilhelm, B., Adler, M., Schaeffel, F., Wilhelm, H.: Mathematical procedures in data recording and processing of pupillary fatigue waves. Vis. Res. **38**, 2889–2896 (1998)
22. Markwell, E.L., Feigl, B., Zele, A.J.: Intrinsically photosensitive melanopsin retinal ganglion cell contributions to the pupillary light reflex and circadian rhythm. Clin. Exp. Optom. **93**(3), 137–149 (2010)
23. Mingzhe, J.: Toukeiteki tekisuto kaiseki 15. In: ESTRELA (182), pp. 44–49 (2009)
24. Mingzhe, J., Murakami, M.: Authorship identification using random forests. In: Proceedings of the Institute of Statistical Mathematics, vol. 55, pp. 255–268 (2007)
25. Morgia, C.L., Ross-Cisneros, F.N., Hannibal, J., Montagna, P., Sadun, A.A.: Melanopsin-expressing retinal ganglion cells: implications for human diseases. Vis. Res. **51**, 296–302 (2011)
26. Morishita, I., Kobatake, H.: Signal Processing (In Japanese). In: The Society of Instrument and Control Engineers, Tokyo, Japan (1982)
27. Nakayama, M., Nowak, W., Ishikawa, H., Asakawa, K.: An assessment procedure involving waveform shapes for pupil light reflex. In: Proceedings of BIOSIGNALS2010, pp. 322–326. INSTICC, New York (2010)
28. Nakayama, M., Nowak, W., Ishikawa, H., Asakawa, K., Ichibe, Y.: Waveform shapes for pupil light responses and their analyzing procedure. In: IEICE Technical Report, pp. 1–6. IEICE, New York (2011)

29. Narita, A., Shrai, K., Kubota, N., Tkayama, R., Tkahashi, Y., Onuki, T., Numakura, C., Kato, M., Hamada, Y., Sakai, N., Ohno, A., Asami, M., Matsushita, S., Hayashi, A., Kumada, T., Fujii, T., Horino, A., Inoue, T., Kuki, I., Asakawa, K., Ishikawa, H., Ohno, K., Nishimura, Y., Tamasaki, A., Maegaki, Y., Ohno, K.: Abnormal pupillary light reflex with chromatic pupillometry in Gaucher disease. Ann. Clin. Transl. Neurol. **1**(2), 135–140 (2014)
30. Olsen, T.W., Feng, X., Kasper, T.J., Rath, P.P., Steuer, E.R.: Fluorescein angiographic lesion type frequency in neovascular age-related macular degeneration. Ophthalmology **111**(2), 250–255 (2004)
31. Park, J.C., Moura, A.L., Raza, A.S., Rhee, D.W., Kardon, R.H., Hood, D.C.: Toward a clinical protocol for assessing rod, cone, and melanopsin contributions to the human pupil response. Invest. Ophthalmol. Vis. Sci. **52**(9), 6624–6635 (2011)
32. Pinkowski, B.: Robust Fourier descriptions for characterizing amplitude-modulated waveform shapes. J. Acoust. Soc. Am. **95**(6), 3419–3423 (1994)
33. Ren, P., Barreto, A., Gao, Y., Adjouadi, M.: Affective assessment by digital processing of the pupil diameter. IEEE Trans. Affect. Comput. **4**(1), 2–14 (2013)
34. Ren, P., Barreto, A., Huang, J., Gao, Y., Ortega, F.R., Adjoudadi, M.: Off-line and on-line stress detection through processing of the pupil diameter signal. Ann. Biomed. Eng. **42**(1), 162–176 (2014)
35. Schimidt, T.M., Chen, S.K., Hatter, S.: Intrinsically photosensitive retinal ganglion cells: many subtypes, diverse functions. Trends Neurosci. **34**(11), 572–580 (2011)
36. Schimidt, T.M., Do, M.T.H., Dacey, D., Lucas, R., Hatter, S., Matynia, A.: Melanopsin-positive intrinsically photosensitive retinal ganglion cells: from form to function. J. Neurosci. **31**(45), 16094–16101 (2011)
37. Stork, D.G.R., Duda, O., Hart, P.E.: Pattern Classification, 2nd edn. Wiley, New York (2001). Japanese translation by M. Onoue, New Technology Communications Co., Ltd., Tokyo, Japan
38. Takane, Y.: Applications of multidimensional scaling in psychometrics. In: C. Rao, S. Sinharay (eds.) Handbook of statistics 26—Psychometrics, pp. 359–400. North-Holland, Amsterdam (2007)
39. TAP Study Group: Photodynamic therapy of subfoveal choroidal neovascularization in age-related macular degeneration with verteporfin. Arch. Ophthalmol. **117**, 1329–1346 (1999)
40. Wilhelm, H., Wilhelm, B., Lüdtke, H.: Pupillography—principles and applications in basic and clinical research. In: J. Kuhlmann, M. Böttcher (eds.) Pupillography: Principles, Methods and Applications, *Clinical Pharmacology*, vol. 18, pp. 1–12. W. Zuckschwerdt Verlag, Munich (1999)
41. Young, R., Kimura, E.: Pupillary correlates of light-evoked melanopsin activity in humans. Vis. Res. **48**, 862–871 (2008)
42. Zahn, C.T., Roskies, R.Z.: Fourier descriptors for plane closed curves. IEEE Trans. Comput. **Vol.C-21**(3), 269–281 (1971)
43. Zhang, D., Lu, G.: A comparative study on shape retrieval using Fourier descriptors with different shape signatures. In: Proceedings of the 5th Asian Conference on Computer Vision, pp. 646–651. Springer, Berlin (2002)

Chapter 2
Prediction of Patients with Alzheimer's Disease Using Extracted Waveform Features of the Pupil Light Reflex in Response to Chromatic Stimuli

Abstract Several analytic approaches were examined in order to estimate the possibility of identifying Alzheimer's patients using features of PLR waveforms from chromatic stimuli. PLRs for three colours of light pulses (red: 635 nm, blue: 470 nm, white: CIE x = 0.28, y = 0.31) at two levels of intensity (10 and 100 cd/m^2) were observed at 60 Hz for 10 s. Pulses consisted of pre-stimulus (2s), light pulse (1 s) and restoration phases (7s). Fifteen features were extracted from each PLR waveform, such as pupil constriction velocity, pupil response delay etc. Participants were seven AD patients (age: 42–84, mean = 68.1) and 12 similar-aged control subjects (age: 62–89, mean = 72.1). There were significant differences in some of the features of PLRs extracted from the two groups (AD and non-AD participants), particularly with the features for blue light stimuli in high brightness, which produced significant reactions in AD patients. The classification performance of using 15 features of the response to blue light stimuli was the highest among responses for all three colours and was higher than the performance using the procedure in the previous study. Two factor scores were extracted from the 15 features across all measurement conditions, and logistic functions were introduced in order to calculate the probability of identifying AD patients. Function parameters were estimated using a Bayesian technique. Least absolute shrinkage and selection operator (LASSO) was applied to sets of PLR features from each light stimulus, together with the ages of subjects, and optimised result sets were obtained. Prediction performance was higher than with the previous procedure. The use of PLRs features from chromatic stimuli for identifying AD was developed and evaluated.

Chapter Co-authors: Wioletta Nowak, Minoru Nakayama, Tomasz Kręcicki, Elżbieta Trypka, Artur Andrzejak, and Andrzej Hacho

Originally published in two papers:
Wioletta Nowak, Minoru Nakayama, Tomasz Krecicki, Elżbieta Trypka, "Analysis for Extracted Features of Pupil Light Reflex to Chromatic Stimuli in Alzheimer's Patients", EAI Endorsed Transactions on Pervasive Health and Technology, 18, e4, 1–10, November 2019.

M. Nakayama, W. Nowak, *Development of Diagnostic Procedures Using Features of Pupil Reactions*, Behaviormetrics: Quantitative Approaches to Human Behavior 21, https://doi.org/10.1007/978-981-95-4502-5_2

2.1 Introduction

The pupil light reflex (PLR), which produces changes in pupil diameter in response to a light pulse of white or green, has been introduced to diagnose Alzheimer's disease (AD) [9, 16]. In addition to this, the recent discovery of intrinsically photosensitive retinal ganglion cells (ipRGCs) [20, 22] reveals the possibility of using various new procedures to diagnose diseases related to retinal conditions by using a shorter light wavelength, such as blue light [14, 29, 33]. For example, PLRs related to ipRGCs can be used to detect symptoms of Age-Related Macular Degeneration (AMD) [39]. Also, the phenomenon has been used to diagnostic procedures for diseases [13, 27, 45]. Additionally, some critical studies have suggested that common sources may be the origin of both AD and AMD diseases [12, 44, 50, 55]. The details of these investigations will be summarised in the next section. However, because most AD patients are elderly, the influence of ageing on PLRs should be evaluated carefully.

The authors have been studying a diagnostic procedure for detecting AD symptoms using PLRs of various types of light pulses and observing the conditions the light pulses produce [43]. Though these results show the possibility of aiding the diagnosis of the disease, a more flexible procedure is required because the prediction performance is insufficient to detect patients with the disease [42]. Though the authors have been examining a procedure to detect AD patients using PLR waveforms, new techniques should be considered to resolve some issues with the existing procedure [42]. In particular, the number of AD patients used in the experimental survey was restricted, and the assessment of the prediction of accuracy was insufficient. Since most participants for the experimental survey are elderly, they may already possess some artefact factors that would affect PLR observations, such as cataracts, or other age-related issues. Of course, a larger number of patients may contribute to accurate estimation of the parameters of the detection procedure, though this increases the number of parameters which need to be estimated, however. The number of samples has been discussed in order to establish a causal relationship.

When the information from the data is insufficient, or only sparse data is obtained, some statistical procedures can be applied to the modelling of the causal relationship [7, 36]. These techniques are introduced to construct a model. Therefore, significant features of PLRs which can be used in the diagnostic procedure should be extracted as needed, even if the number of samples is insufficient. Other factors should be estimated in order to reduce the influence of the problem of

---

Wioletta Nowak, Minoru Nakayama, Tomasz Kręcicki, and Andrzej Hachoł "Detection Procedures for Patients of Alzheimer's Disease Using Waveform Features of Pupil Light Reflex in Response to Chromatic Stimuli", EAI Endorsed Transactions on Pervasive Health and Technology, Vol. 6, Issue 23, pp. e6:1–11, December 2020.

insufficient data mentioned above. If the distribution of features can be estimated, the possibility of diagnosing AD patients may be predicted using a Bayesian process, such as the Markov chain Monte Carlo (MCMC) method [36]. During the analysis, restriction of the distribution of features often influences optimising procedures for the prediction models since the extracted features are distributed sparsely. In order to analyse sparse distributed data samples, some sparse estimation techniques have been developed, such as the least absolute shrinkage and selection operator (LASSO) technique [7, 18, 28], and examination of the effectiveness of this estimation procedure is required.

In this chapter, the problems mentioned above, such as the number of patients or appropriate features of PLRs, are discussed using several methods of analysis of observed PLR responses.

First, the possibility of detecting Alzheimer's disease in patients using a set of features of PLRs in response to chromatic stimuli is examined. The PLRs of both AD patients and healthy control participants were observed using stimuli consisting of white, red and blue light to activate ipRGCs and conventional retinal ganglion cells.

Second, features of PLRs are extracted and the differences between AD patients and control group subjects are discussed, based on the results in the previous study [42]. Some procedures for predicting the probability of diagnosing AD patients are compared [41]. In addition, the sparse estimation technique is also applied to this prediction in order to improve its accuracy. The following topics are addressed:

1. The features of PLRs for chromatic stimuli shown at two brightness levels to healthy and AD patients are compared.
2. The influence of a participant's age on PLRs in response to the stimuli is measured and analysed.
3. The possibility of detecting AD in patients using the features of PLRs is examined. The contributing features are also extracted.
4. Features of PLRs are analysed, and their factors are extracted. The differences in features and factor scores between AD patients and control group subjects are compared. Also, the influence of ageing is evaluated.
5. Logistic regression is introduced to calculate the probability of detecting the disease in AD patients and control group subjects. The models of fitness of the groups are then compared.
6. The MCMC technique is introduced to estimate the parameters of the models, and the performance of the models is discussed.
7. The LASSO technique is introduced to improve prediction performance, and selected features are discussed.

As noted above, some diagnostic procedures based on PLR waveforms have been developed, and their diagnostic possibilities and performance have been examined.

2.2 Related Works

2.2.1 *Alzheimer's Disease (AD)*

Clinical features of Alzheimer's disease (AD) include progressive cognitive decline, affecting memory, learning, language, deterioration of sleep and disturbances of normal circadian rhythms [48, 53]. Alzheimer's disease affects approximately 10% of people aged 65 or over. It is a chronic, progressive neurodegenerative disease characterised by changes in many cognitive and behavioural aspects that significantly interfere with function in everyday life.

One model of AD formation is that many factors combine to cause the accumulation of amyloid β (Aβ) in the brain, which in turn causes synaptic dysfunction, tangle formation and neuronal death. These pathology conditions are found in the central nervous system, as well as retina (especially mRGC) [34] and in the locus coeruleus (LC) [24, 38, 49].

There is widespread agreement on the need for identification of an early stage of AD using minimally invasive (or even noninvasive) inexpensive screening test. In addition to the debilitating symptoms endured by AD patients, this disease imposes a huge social and economic load on the society.

In order to diagnose AD patients and persons presenting symptoms of dementia, such as those with mild cognitive impairment (MCI) [21], the MMSE (Mini-Mental State Examination) question inventory [15] is often used during clinical consultations. As MMSE is used by observers to mark conditions during subject observation, the validity of the results depends on the activity of the subject during the examination. Therefore, some revisions have been introduced to improve the accuracy of consultations [52]. Also, more objective indices of subject's responses have been required in order to help medical doctors during clinical consultations.

2.2.2 *PLRs with AD Patients*

PLRs have been used with AD patients, and the influences of light intensity [47], light pulsing [25] and the resulting behaviours [17, 19] have been examined using the response to some forms of light. In these studies, the characteristics of PLRs of AD patients, comparable AD-like participants such as people with mild cognitive impairment (MCI) [21] and healthy age-matched control subjects were compared. Since PLRs of elderly peoples are influenced by factors concerning age, dynamical features of pupil responses such as amplitude, latency and others are affected [8]. This is the reason for the requirement of preparing age-matched subjects. Though there are some differences between AD patients and other subjects [19], an appropriate extraction procedure for AD patients that considers the factor of age needs to be created.

2.2.3 PLRs Based on ipRGCs

The initial research into diagnosing AD patient was conducted before ipRGCs were discovered, unfortunately. As many recent studies have indicated, another type of PLR which responds to the melanopsin-based ipRGCs has been discovered [11, 23]. This phenomenon can be used as a disease-detecting diagnostic procedure [29, 33], since the PLR behaviours of ipRGCs are different from conventional reactions which are based on the rod and cone system [45, 56]. These detailed differences in PLR mechanisms in response to chromatic stimuli at several brightness levels have been discussed [32, 45]. In addition to the behavioural differences of ganglion cells, ipRGCs are scattered across the retina, while the conventional ganglion cells, such as the rod and cone system, are located in specific areas. By using these differences in PLR behaviour, PLRs in response to chromatic stimuli have been observed in patients with retinal diseases such as Pigmentosa or Glaucoma [13, 14, 27]. Also, this phenomenon was used with AMD patients, and a diagnostic procedure has also been discussed [39]. The patients do not recognise the influence of AMD when the diseased area appears at the periphery of the retina. In this case, the cone-based PLRs may not be influenced, but ipRGC-based PLRs will be influenced in regard to the difference in their distribution. Therefore, the differences in the two types of PLRs may suggest a retinal condition as the possible reason for the size of the damaged area caused by AMD. To examined the differences, several features of waveform shapes of PLRs in response to red and blue light stimuli were compared using discrimination analysis [39].

2.2.4 Possibility of Diagnose Using PLRs with ipRGCs

The previous studies revealed some significant differences in features of PLRs between AD patients and healthy age-matched control healthy subjects. Modern day statistical analysis offers the possibility of identifying potential patients. To obtain accurate results, appropriate feature extraction and a robust discriminant technique are required.

While some diagnostic procedures have been developed using ipRGC-based PLRs, conventional PLRs in response to white lights [16] and green lights (wavelength = 585 nm) [9] are still used to diagnose AD patients when the influence of age is not a consideration.

These studies are also examining whether typical features of PLRs can be biomarkers for AD patients [19, 21, 31]. In the case of AMD patients, the retinal conditions influence PLR waveforms, in particular the differences between blue and red light stimuli. Recently, in addition to amyloid β accumulation, retinal ganglion cell and nerve fibre layer loss [54], retinal bio-markers such as retinal vascular conditions can now be used as indices of symptoms for AD patients.

Interestingly, the early stages of AD diseases show histopathological changes in the structure of one of the types of RGCs, namely ipRGCs. In addition, it is very interesting that in AD, ipRGCs are the first to be destroyed even when the remaining RGCs remain pathologically unchanged.

Current AD patient studies suggest that the presence of amyloid β (Aβ) influences the disease [37, 51] and also that this material, in particular, the retinal pigment epithelium (RPE), is a common risk factor for AMD [55]. Using these two approaches, the detailed mechanisms of both diseases have been discussed by medical researchers [12, 44, 50]. The results have shown that due to the condition of the retinal, PLR responses may reflect both AMD and AD.

Therefore, a diagnostic procedure that has been developed for AMD can be used as the common diagnostic procedure for detecting symptoms of AD using PLR waveform shapes. This is a quick and simple procedure. The performance of preliminary procedures has produced results that are insufficient [43]. An improved procedure using ipRGC-based PLRs should be developed, and the diagnostic performance examined together with conventional methods. In particular, PLR responses have been analysed in order to detect symptoms of AD, since ipRGCs and ipRGC-based PLRs have been discovered [10, 57]. However, the diagnostic procedure using PLRs for AD patients continues to examine patients using observation, though some clinical procedures have been introduced [10].

2.2.5 Predictions of AD Patients

Computational diagnostic procedures for detecting AD patients or patients with dementia have been studied using human behavioural features. Unfortunately, the typical features that present AD symptoms are not currently being examined. Different information from human actions can be applied to the observation and prediction of cognitive performance using machine learning techniques. A summary of related works is listed in Table 2.1.

Since speech is observed by medical doctors during consultations, acoustic and lexical features are employed as a diagnostic procedure for a small scale set of data (nine AD patients and nine control participants) [26]. The sound of a patient's speech can be recorded during medical consultation, and several projects have begun to summarise large sets of speech corpus data. Research groups can conduct computational prediction and classification of the level of diseases in participants using a corpus consisting of over 100 AD patients [30, 59]. Performance depends on the data sets and the prediction techniques, and it does not seem easy to compare the accuracy of the two proposed procedures with previous studies.

Another approach is to observe human gait behaviour [6]. Since we need to pay a lot of attention to our bodies while we are walking, some interactions may affect this behaviour, causing our attention to be distracted. Recently, image processing techniques that can easily extract gait features are being applied to diagnosing

Table 2.1 Summary of the related works

	MMSE	Acoustic features + ML [26]	Lexical analysis + ML [30]	Gait features + ML [6]	ipRGC base PLRs + ML [42]
Data feature	Responses to question items	Acoustic features of speech (3–5 min)	Lexical analysis for conversation audio	Gait features of walking with dual tasks	Extracting features of PLRs
Data sample	MD diagnostic	AD:9 + control:9	AD:194 + control:98	103 MCI participants	AD:19 + control:12
Attribution	Clinical observation	Age, sex, education, MMSE score	Corpus base analysis (546 values)	Task performance, MMSE score, age	Age, sex
Technique	Clinical diagnosis	MLP	ANN	SVM	RandomForest
Accuracy	N/A	92.0 %	84.4 %, AUC=0.92	AUC=0.747	78.4 % (AD:57%,Control:90%)

elderly people. The possibility of dementia in 103 elderly participants was predicted using a machine learning technique [6].

As data deviation grows when the number of participants increases, AD patient prediction performance or prediction of the degree of development of the disease, including prediction using features of PLRs, does not necessarily improve through the use of larger-sized sets of data. In particular, feature extraction or feature assessment for making prediction is still preferred, before adaptations that would increase the size of data set are made.

Also, features of PLRs have been applied to the prediction of AD patients using conventional machine learning techniques [42]. AD patient prediction performance should also be improved using other metrics from the previous studies, though Table 2.1 shows an acceptable level of performance for predicting control subjects. As AD patients possess various stages of the disease, the performance of binary classification using categories such as "patient" vs. "healthy" is sometimes not useful for the diagnosis, and an intermediate assessment such as the level of MCI may be required. For this approach, the probability of diagnosing AD patients would help participants to understand the progress of the disease.

Therefore, more appropriate feature selection and a more appropriate prediction procedure should be considered, using the data sets for PLRs with various chromatic stimuli [42]. The following sections introduce new techniques in order to improve performance.

2.3 Method

2.3.1 Participants

A conventional PLR experiment was performed using 19 participants (42∼89 years old, mean age:70.6), 12 of which were healthy individuals with normal vision (the control group: 62∼89 years old, mean age: 72.1) and 7 who were patients with Alzheimer's disease (AD patients: 42∼84 years old, mean age: 68.1) who had already been diagnosed by medical doctors. It was not easy to invite volunteers who were aged over 80. The age levels are summarised in Table 2.2.

The experiment confirmed by the local ethical committee of the Wrocław Medical University which was established in accordance with the Declaration of Helsinki. Patients and healthy controls gave informed written consent on paper before the experiment and debriefed afterwards about the purpose of the experiment.

Table 2.2 Number of subjects by age

Label	Age	Control	Patient
L	≤ 70	5	3
M	71∼80	5	3
H	$81 \leq$	2	1

2.3.2 PLR Measurement

The stimuli consisted of three chromatic lights, red (635 nm), blue (470 nm) and white (CIE x: 0.28, y: 0.31), at two levels of brightness (10 and 100 cd/m^2). These stimuli were labelled as r10, r100, b10, b100, w10 and w100.

The duration of observations was 10 seconds, with the first 2 s being a pre-stimulus phase as a rest period, followed by a 1 s light pulse and 7 s as a restoration phase. Pupil diameters were measured in mm at 60 Hz using a system developed by some of the authors, which is illustrated in Fig. 2.1 [40]. Figure 2.1 shows a diagram of the measuring unit, which is installed on an ophthalmological slit lamp base as the image acquisition module, and it consists of the infrared (IR) illumination, the optical path and a single camera with a frame grabber. The illumination part consists of four IR light-emitting diodes (LEDs), two for the left eye (OS)(IR_OS) and two for the right eye (OD)(IR_OD). The IR light is reflected by the corneal surface, transmitted by the optical paths, and projected onto a high-speed high-resolution camera. The PC stores the acquired data as a sequence of grey-scale pupil images in bmp format. The image processing module estimates the measured pupil parameters frame by frame, individually for the left and the right eyes. Finally, the obtained pupil dynamic waveforms are stored. PLRs for each stimulus were observed in single trials using a repeated-measure design.

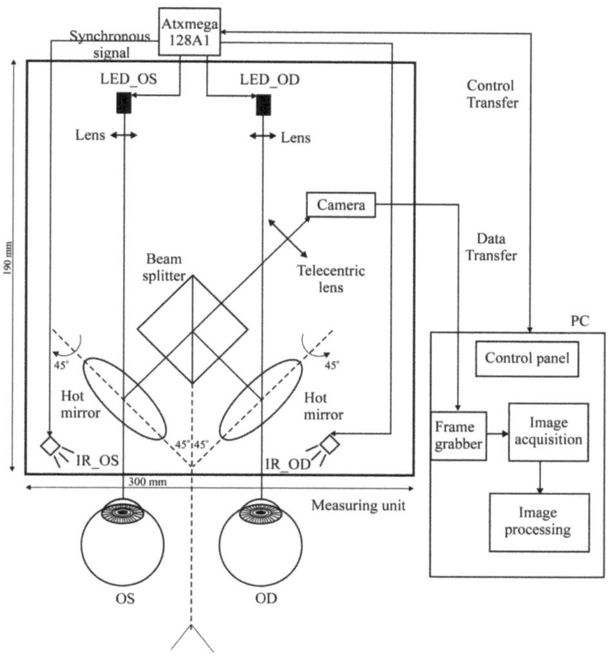

Fig. 2.1 Diagram of measuring equipment [40]

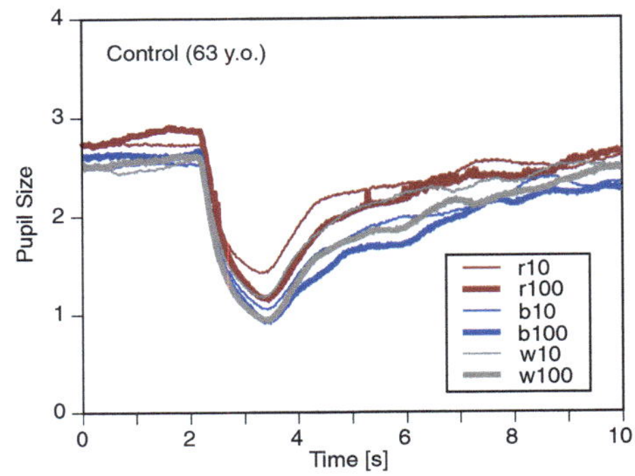

Fig. 2.2 An example of PLRs for a control group subject

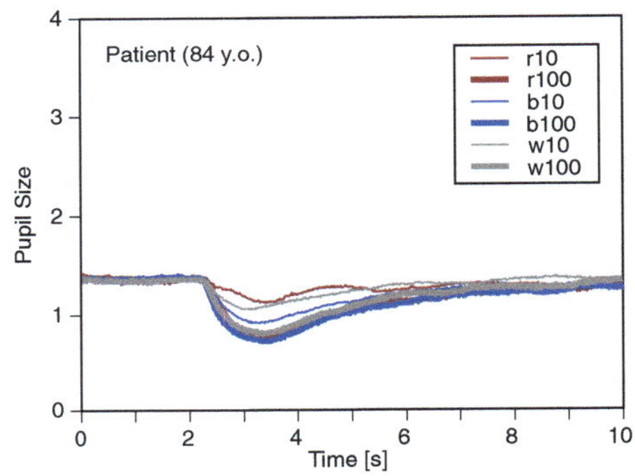

Fig. 2.3 An example of PLRs for an AD patient

Examples of measurements for a healthy individual are shown in Fig. 2.2 and for an AD patient in Fig. 2.3. In these figures, PLRs are illustrated in response to six stimuli, namely the three colours and two levels of brightness.

2.4 Feature Extractions from PLRs

In comparing the waveforms of PLRs between two Figs. 2.2 and 2.3, the information from the PLRs may represent some of the features of AD patients. Two types of feature representations need to be considered in order to extract the specific reactions of the two approaches.

2.4.1 Fourier Descriptors for PLR Waveform Shapes

PLR's waveform shapes can be represented using Discrete Fourier Transform (DFT) [46, 58]. Discrete Fourier Transform (DFT) was applied to all responses. The signal $x(n)$ can be noted as an Eq. (2.1) using DFT [35].

$$x(n) = a_0 + \sum_{k=1}^{N/2} \left(a(k)cos\left(2\pi k \frac{t(n)}{N\Delta}\right) + b(k)sin\left(2\pi k \frac{t(n)}{N\Delta}\right)\right) \tag{2.1}$$

$$a_0 = X(1)/N$$

$$a(k) = 2\,real(X(k+1))/N$$

$$b(k) = 2\,imag(X(k+1))/N$$

Some of the magnitudes of the coefficients in the early orders of the series, $a(k)$ and $b(k)$, and the DC component (a_0) were employed to form feature vectors of the waveform shapes [39]. The magnitudes of the coefficients, including a_0, $FD_i(i = 0, \ldots, 5)$, can be noted as follows:

$$f = [FD_0, FD_1, \ldots, FD_5] \tag{2.2}$$

The features are affected by individual factors, so that a standardised feature using the component, FD_1, which is converted from the above vector as follows [58]:

$$f = \left[\frac{FD_2}{FD_1}, \frac{FD_3}{FD_1}, \ldots, \frac{FD_5}{FD_1}\right] \tag{2.3}$$

As a result, the features of the waveform shapes can be noted as a vector consisting of five components, including the DC component.

2.4.2 PLR Feature Definitions

A typical PLR waveform shape is illustrated in Fig. 2.4. In the figure, the light pulse overlaps for a period of 2~3 seconds. As the figure illustrates, there are pupillary response delays due to the shrinking of pupil and its restoration to the normal size.

Some features that are extracted to specify the PLR response, and these variable features are summarised in Table 2.3. They are pupil size, velocity of pupillary change, duration of change and integration of the waveform. These features are calculated for each PLR response.

2.4.3 Prediction Procedure

Regarding the differences between Figs. 2.2 and 2.3, PLRs may be influenced by disease in patients with AD. This possibility is examined by introducing classification analysis using a Random Forest technique.

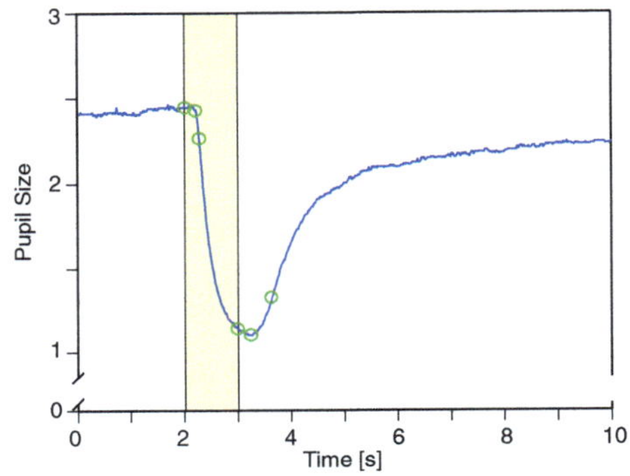

Fig. 2.4 Feature extraction from PLR responses

Table 2.3 Definitions of PLR features

Variable	Definition and notes
ps_base	Mean of the pupil size for time before light pulse
ps_lon	Pupil size where light pulse on
ps_loff	Pupil size where light pulse off
ps_min	Minimum pupil size
RA	Range of pupil size (ps_lon - ps_min)
v_con	Max velocity of pupil constriction stage during light pulse on
v_rest	Max velocity of pupil re-constriction stage after light pulse off
ac_max	Max acceleration for pupil re-constriction during the constriction
t_delay	Pupil response delay
t_min	Time when "ps_min" appears
t_v_con	Time when "v_con" appears
t_v_rest	Time when "v_rest" appears
int_con	Integration of constriction phase
int_rest	Integration of restoration phase
int	Overall integration (int_con + int_rest)

The number of variables used to train the model for the data sets was considered carefully. The number of variables sampled was optimised using the tuning function of the "randomForest" package for R [2].

The results of a two-class classification were evaluated using a Random Forest technique, which consists of training the model using a part of the assigned data set, and conducting classification prediction using the model and the remaining data. The results produced an average of ten training and prediction trials, in order to evaluate classification performance. Prediction performance was evaluated using error rates for each class of healthy and patient participants and for the overall error rate.

2.5 Results of Classification Using PLR Waveform Shapes

2.5.1 *Features Comparisons in PLRs*

Features of PLRs are defined in the previous section. These features are compared between the two groups of healthy control and AD patient participants.

2.5.1.1 Features of Waveform Shapes

The features of PLR waveform shapes for all PLRs throughout the 10 s observation period were extracted using the procedure mentioned in the above section.

Since the waveform shapes in Figs. 2.2 and 2.3 looked different, the features were compared. The major features are the DC components such as the overall magnitude of the pupillary changes. Figure 2.5 shows the means of the DC components for the three colours at the two levels of brightness (10 and 100 cd/m^2).

The results confirm that the mean amplitudes for blue lights are the highest and that the mean amplitudes decrease in the order of white to red light. Also, the brightness of the stimuli affects the mean amplitudes of all colour conditions independently. In a comparison of the means of the two groups of healthy control and AD patients, there are no significant differences. The overall amplitudes of waveform shapes are comparable between the two groups. This result coincides with the previous study [43].

The four-dimensional components were produced by using the extraction procedure for the waveform shapes. Though each component of every stimulus condition was compared between the two groups, there were no significant differences except for one feature in the b10 condition. This result suggests that AD patients cannot be detected using the differences of the features.

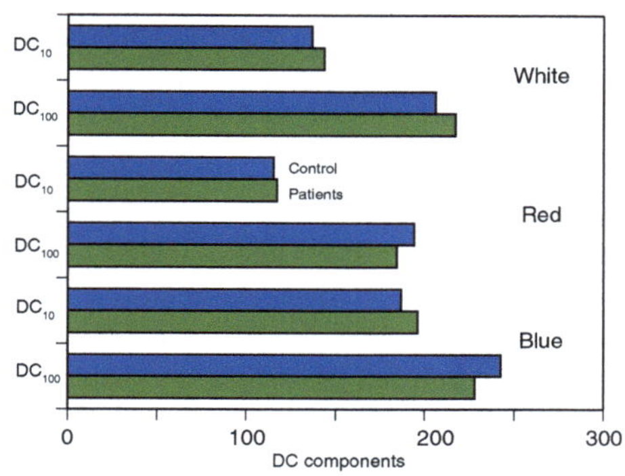

Fig. 2.5 Comparison of DC components between stimulus conditions

2.5.1.2 Comparison of Extracted Features Between the Two Groups

The extracted features from each stimulus were compared between the two groups. The results are summarised in Table 2.4. A statistical test, known as a t-test, was applied to these values. When there was a significant difference ($p < 0.05$) between pairs of values, the values are displayed in bold face. As the table shows, there are many significant pairs for the b100 and r10 conditions, but few significant pairs for the white stimulus. In regard to the significant differences for the b100 condition, such as pupil size, velocity and acceleration of pupillary change, the pupil size for AD patients is relatively small and responds slowly.

All participants are elderly, and in addition to being AD patients, their ages may affect pupil responses. The effect of two factors (participant group and age level) on pupillary changes was examined using two-way ANOVA. The variables with deviations which contribute most significantly were selected ($p < 0.05$) and are summarised in Table 2.5 using means across age levels. These means change along with age levels. Most variables selected are related to velocity and time delay. Since there are few significant interactions between the two factors, there are no interactions between two factors in regard to features of PLRs. In addition, most significant differences between age levels appeared for white stimulus and most differences between the two groups occurred when w100 light was used. These results may be related to the mechanism of the PLR, and thus a detailed analysis of this will be a topic of our further study.

2.5.2 Prediction Performance

2.5.2.1 Prediction Using Features of Waveform Shapes of PLRs

The possibility of extracting patients with AD was examined using five-dimensional features of waveform shapes for a stimulus condition [43]. Some data sets were prepared as combinations of features for stimulus condition consisting of a combination of features for red, blue and white colours, at two levels of brightness (10 cd/m^2 and 100 cd/m^2).

The combinations were all conditions for red and blue light stimuli at two levels of brightness (red+blue), all conditions for the three light stimuli at two levels of brightness (r+b+w) and selected combinations of the three conditions such as [r10, b10, b100] (rbw). As mentioned in the Method section, a Random Forest technique, which is based on a bootstrap method [36], was applied to the data sets as the number of samples is quite limited. Performance was summarised using mean correct ratios of the overall tenfold cross-validations for patients and for healthy control groups. The results are indicated in the upper panel of Fig. 2.6. The overall discrimination performance is around 70$\sim$80%, and for the selected feature combinations it is slightly higher than for the other two conditions. Prediction performance for AD patients stays at lower levels though the levels for healthy

Table 2.4 Means of PLR features

Feature	b10(N=19)		b100(N=17)		r10(N=18)		r100(N=19)		w10(N=19)		w100(N=19)	
Variable	Control	Patient	Control	Patient	Control	Patient	Control	Patient	Control	Patient	Control	Patient
ps_base	19.88	16.69	**20.51**	**14.21**	**20.16**	**14.37**	19.59	16.41	19.2	16.40	19.20	16.46
ps_lon	20.24	16.72	**20.81**	**14.18**	**20.29**	**14.22**	20.00	16.26	19.56	16.54	19.45	16.46
ps_loff	10.90	8.02	**9.76**	**6.76**	13.02	9.29	10.15	7.68	11.61	8.80	9.14	7.03
ps_min	10.68	7.59	**9.20**	**5.88**	12.75	8.96	9.76	7.21	11.31	8.56	8.77	6.29
RA	9.56	9.13	**11.60**	**8.30**	7.53	5.27	10.24	9.05	8.26	8.08	10.69	10.17
v_con	−0.46	−0.47	**−0.51**	**−0.31**	**−0.36**	**−0.22**	−0.44	−0.37	−0.40	−0.40	−0.48	−0.42
v_rest	0.13	0.10	**0.18**	**0.08**	**0.13**	**0.08**	0.16	0.10	0.14	0.10	0.14	0.11
ac_max	−0.06	−0.07	**−0.07**	**−0.04**	**−0.05**	**−0.03**	−0.06	−0.05	−0.06	−0.05	−0.07	−0.06
t_delay	0.24	0.26	0.23	0.24	0.25	0.26	0.24	0.25	0.26	0.28	0.23	0.23
t_min	**1.13**	**1.31**	**1.30**	**1.43**	1.09	1.26	1.22	1.36	1.10	1.22	**1.26**	**1.41**
t_v_con	0.34	0.35	**0.33**	**0.36**	0.35	0.39	0.37	0.37	0.37	0.38	0.32	0.33
t_v_rest	1.84	1.78	1.91	2.14	1.65	2.06	1.88	2.11	1.67	1.96	1.74	1.98
int_con	293.1	277.0	**345.0**	**211.0**	**230.0**	**144.5**	301.5	260	249.4	241.6	330.5	300.3
int_rest	748.5	823.0	1015.9	802.7	532.1	444.1	849.8	796.9	612.6	652.3	906.9	941.7
int	1041.6	1100	1361	1014	762.1	588.7	1151.3	1056.9	862.0	893.9	1237.3	1242.0

Table 2.5 Age affected features (list of significant variables ($p < 0.05$))

Stimulus	Variable	Age levels		
		L	M	H
b10	ac_max	−0.11	−0.05	−0.04
b100	ps_min	8.0	6.9	10.9
r10	t_v_con	0.37	0.36	0.43
r100	t_delay	0.23	0.24	0.31
	int_con	381.4	249.5	176.3
w10	RA	11.2	7.9	4.7
	v_con	−0.64	−0.33	−0.25
	ac_max	−0.09	−0.04	−0.03
w100	RA	14.3	9.6	7.6
	v_con	−0.64	−0.39	−0.33
	t_v_con	0.28	0.34	0.36
	int_con	459.6	280.8	217.9
	int_rest	1279.1	853.7	659.8
	int	1738.8	1134.5	877.7

Age factor is significant ($p < 0.05$)

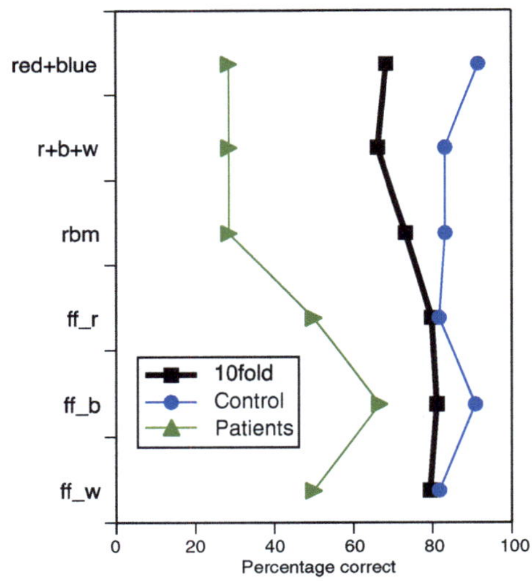

Fig. 2.6 Comparison of classification performances

control group for controlled healthy group are over 80%. The overall prediction performance is comparable with the previous preliminary experiment where the duration of light pulse was 10 seconds. This evidence shows that the prediction performance between two durations of light pulses (10 seconds and 1 second) is comparable. In both cases, the prediction performance for AD patients is quite low, and thus additional improvement is necessary.

2.5.2.2 Prediction Using Features of PLRs

The features of PLRs defined in Table 2.3 were used in the prediction and so was the Random Forest technique. To evaluate the effectiveness of stimulus colours, prediction performance was compared using a set of one of the three colours in two levels of brightness (10 cd/m^2 and 100 cd/m^2). This performance is summarised in Fig. 2.6 for three colour conditions (ff_r: red colour stimuli, ff_b: blue colour stimuli, and ff_w: white colour stimuli). The performances using features of PLRs improved the percentage of correct for AD patients. The tenfold cross-validations of the overall performance are higher than are the ones that use the features of waveform shapes. Though the percentage correct predictions for AD patients increased, it remained at the level of chance. The performance (tenfold) for blue stimuli (ff_b) is the highest of the six conditions. Also, the percentages for both the control and AD patients are the highest. As shown in Table 2.4, there are significant differences between AD patients and the healthy control group. The feature values for blue lights may affect prediction performance. In addition, the performance using the b100 condition is at the same level as the performance when the two levels of blue light (b10 and b100 conditions) are employed (see Table 2.7).

To present the actual prediction performance, 19 participants are placed in clusters using the PLR features of the b100 condition. Many significant differences were observed, as Table 2.4 shows. Hierarchical cluster analysis using an averaging method was applied to features of the b100 condition, and a dendrogram of the results was created. The dendrogram is illustrated in Fig. 2.7. The labels in the figure show the participants and their ages, and the subjects marked with an "#" are AD patients. The lower cluster includes five of the six patients, but three healthy

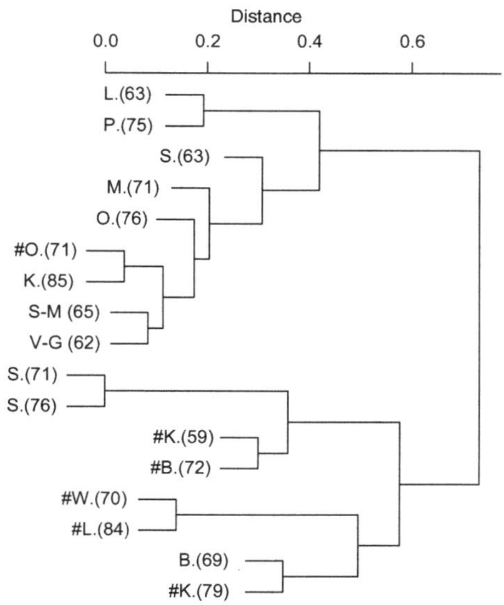

Fig. 2.7 Dendrogram of participants using features of b100 condition; # marks are patients

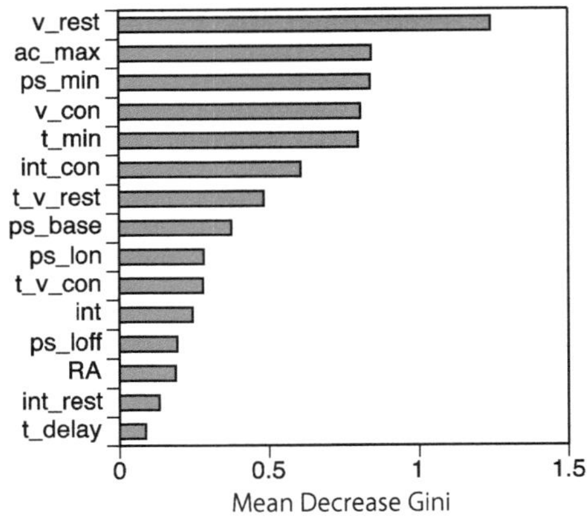

Fig. 2.8 Contribution weights in features of b100 condition for the prediction

participants are also in the same cluster. It is not easy to predict perfectly which patients are which, but the results show the possibility of extracting AD patients using the performance.

To measure the degree of contributions of features of PLRs, the level of importance of the variables extracted for prediction in the b100 condition is summarised in Fig. 2.8. This figure shows the importance of the variables that were introduced to the prediction using the Random Forest method. The largest contribution is v_rest (maximum amplitude of pupil re-constriction velocity). As following the order of the degree of contribution, v_con (maximum amplitude of pupil constriction velocity), ac_max (minimum amplitude of pupil acceleration), ps_min (minimum pupil size) and its time t_min are confirmed their contribution. Most variables represent pupil constriction behaviour.

As these features are based on blue light stimulus at a high level of brightness, the pupil response of ipRGCs-based reactions may provide information about patients. Further extraction of typical features may improve prediction performance. However, as it is not easy to gather a large enough sample of patients, other types of prediction procedures should be considered. They will be the subjects of our further study.

2.5.3 Discussion

The values of features of PLRs defined above were compared between AD patients and healthy control subjects, as shown in Table 2.4. The results of t-tests show that the most significant differences appeared when b100 (blue 100 cd/m^2) and r10 (red 10 cd/m^2) were used. There were few differences when white light stimulus was used. In particular, significant differences in pupil diameters were observed for light pulse, speeds and acceleration of the change in response to b100 stimulus. For AD patients, this result suggests that pupil diameters under light pulse and their change in speed are relatively small, and the responses are delayed.

As mentioned above, both being an AD patient and factors related to ageing may influence pupil response. An analysis of the factors related to ageing was conducted on the features of PLRs, and the results are summarised in Table 2.5. The results show the gradual influence of age in accordance with the age levels (L, M, H). Significant differences ($p < 0.05$) are observed in v_con (maximum amplitude of pupil constriction velocity), and their delays such as t_v_con (time when "v_con" appears), t_delay (pupil response delay). The differences may have occurred because the size of the pupil constriction is relatively small. In addition, most factors of interactions between ageing factors and the two groups of subjects (AD patients and the healthy control group) were not significant. Both factors operate independently of each other.

As mentioned above, for white light pulse there are significant differences in ageing factors for features of PLRs, although significant differences between the two groups of subjects appeared when blue light was used. PLRs for blue light which result mainly from ipRGCs can be used to evaluate the ageing factor independently. This kind of response behaviour may be caused by a mechanism in PLRs which concerns ipRGCs and conventional rod-cone systems.

Prediction performance using the PLR features extracted was better than the performance using features of PLR waveform shapes. In particular, the performance using features of PLRs in response to blue light stimuli (10 cd/m^2, 100 cd/m^2) was the highest. Once again, the pupil responses for blue light, such as PLRs based on ipRGCs, reflect some of the features specific to AD patients. Though prediction performance was not perfect, as Fig. 2.7 shows, it can be used as a reference.

In a detailed analysis of prediction performance using the blue light stimulus b100, the importance of features of PLRs was evaluated, as Fig. 2.8 shows. As the contributing variables suggest, a more sophisticated prediction procedure needs to be developed. This will be a subject of our further study.

In addition to the above discussion, the number of participants in this study might be small scale comparing with the general classification tasks. On the other hand, there are many deviations on the features of elderly people or AD patients; then the larger sample size may not resolve the individual difference issues. During the assessment for classification performances, tenfold validation index is employed in order to consider validation of the procedure. If there were insufficient estimations for the trained classifiers, they can be improved step by step with sample size growing. Therefore, more precise models will be developed along with our study progress. More detailed discussion will be a subject of our further study.

2.6 Results of Prediction Procedures Which Have Been Developed

2.6.1 Characteristics of Features in Response to Stimuli

2.6.1.1 Comparison of Features

The features extracted for each stimulus were compared between the two groups. The results are summarised in Table 2.4. When there is a significant difference between pairs of values, the values are displayed in bold face. As the table shows, there are many significant pairs for the b100 and r10 conditions, but few pairs for the white stimulus. In regard to the significant differences for the b100 condition, such as pupil size, velocity and acceleration of pupillary change, the pupil size for AD patients is relatively small and responds slowly.

All participants are elderly, and in addition to being AD patients, their ages may affect pupil responses. Therefore, the effect of the two factors (participant group and age level) on pupillary changes is examined using two-way ANOVA. These means change along with age levels. Most variables selected are related to velocity and time delay. Since there are few significant interactions between the two factors, they may be independent of each other in regard to PLR features.

2.6.1.2 Factor Analysis

There are significant differences in some of the PLR features between the AD and control groups. Though their variables exhibit the qualitative tendencies of a pupillary change, the sources could not be determined, as the physical variables and measurement units are completely different; some are expressed using sizes, and others are expressed as velocities or accelerations.

Factor analysis was used to extract latent sources of the variables, which were measured repeatedly in order to reduce the number of dimensions of the features of PLRs. As the number of features is too great for all to be evaluated, some features are correlated with others. In general, the factors extracted represent the variations. Since overall integration (*int* in Table 2.4) is a summation of two parts, such as int_con and int_rest, have been omitted during the analysis, thus the remained 14 variables were measured. A two-factor structure was estimated using a principal component solution and a screw plot. A factor loading matrix using Promax rotation was produced, as shown in Table 2.6.

The fundamental variables of each participant, such as pupil size, commonly contribute to both factors. The second factor contains variables that are concerned with features of the progress of restoration of the pupil after a pulse of light, and the first factor contains the remaining variables of the features of PLR. As mentioned above, both factors contain two variables, and there is a significant correlation between these two factors ($r = 0.38$). Since even the contribution ratio of each

Table 2.6 Factor loading matrix for PLR features with Promax rotation

Variables	Factor 1	Factor 2
ps_base	**0.596**	**0.586**
ps_lon	**0.596**	**0.586**
int_rest	**1.007**	−0.271
RA	**0.997**	−0.064
int_con	**0.988**	0.006
v_con	**−0.906**	−0.067
ac_max	**−0.902**	−0.021
t_v_con	**−0.737**	0.110
t_delay	**−0.712**	0.062
v_rest	**0.643**	0.099
ps_min	−0.111	**1.025**
ps_loff	−0.079	**0.999**
t_min	0.086	**−0.643**
t_v_rest	0.066	**−0.438**
Contribution ratio (1)	0.42	0.21
Contribution ratio (2)	0.52	0.32
Correlation between factors		$r = 0.38$

(1): each factor with other factors eliminated
(2): each factor with other factors ignored

factor when the other factors are eliminated is over 60%, the two factors can account for the deviation of features of PLRs.

The factor scores ($factor$ 1 and $factor$ 2) were calculated using the factor loading matrix, and the means for each stimulus are summarised in Fig. 2.9, according to group. The horizontal axis represents the first factor, and the vertical axis represents the second factor. The error bars show standard errors. The two groups are indicated using suffixes, such as "p" for patients and "c" for the control group.

When means for red ($r10$ and $r100$) or white ($w10$ and $w100$) light stimulus between the two groups are compared, they are located proximate to each other, but the means for blue ($b10$ and $b100$) light stimulus of the two groups are relatively far apart from each other. Pupil reactions in response to light colours are represented in Fig. 2.9. In the results of t-tests applied to pairs of means for the two subject groups, there are significant differences in the first factor scores for b100 ($t(15) = 2.64$, $p < 0.05$), in the second factor scores for b100($t(15) = 2.88$, $p < 0.05$) and for w100($t(17) = 2.15$, $p < 0.05$). Also, differences with little significance were observed for the second factor scores for b10($t(17) = 2.07$, $p < 0.10$) and for r10($t(17) = 1.77$, $p < 0.10$).

The results show that between the two subject groups there are significant differences in both factor scores for the b100 condition. As Table 2.4 shows many significant differences, the factor scores also represent these differences. The second factor seems to reflect the difference in features of PLRs between the two groups.

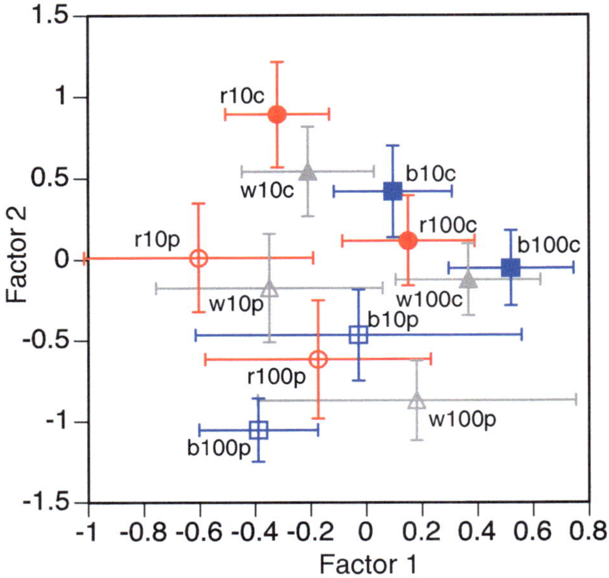

Fig. 2.9 Distribution of factor scores with error bars showing standard errors

Although the influence of age level on factor scores was analysed, it did not affect either factor.

2.6.2 *Introducing Logistic Regression*

2.6.2.1 Function Optimisation

As mentioned in the Introduction section, this work introduces the probability of diagnosing AD patients in order to estimate the progress of the disease using logistic regression analysis to predict binary response variables (p) and features of PLRs. Here, $p = 1$ for the control subject and $p = 0$ for the AD patient, so that p is given using the following equation with logit function:

$$\hat{y}_i = a + b_1 * factor1_{i,j} + b_2 * factor2_{i,j}$$

$$p_i = \text{logit}^{-1}(\hat{y}_i) = \frac{1}{1 + \exp(-\hat{y}_i)}$$

Suffix i represents the subject, and suffix j represents the light stimulus condition.

Logistic regression analysis was applied to the above factor scores for several conditions, such as the two factor scores for a specific light stimulus condition (light intensity: 10 or 100 cd/m^2), the four factor scores for one colour condition (blue, red or white, including light intensities 10 and 100 cd/m^2) and further combinations such as two colours or all conditions (six conditions and two factor scores). Every model was evaluated for fitness of model using AIC (Akaike Information Criteria) and for prediction accuracy using R^2. During validation of parameters, the likelihood or fitness of the prediction is evaluated using AIC or other statistical indices in addition to the level of accuracy. Accuracy was measured using an appropriate threshold, and performance was summarised using two-dimensional metrics such as true and false positives. The relationships are then illustrated as ROC (Receiver Operating Characteristic) curves. Figure 2.10 shows ROC curves for every stimulus condition. The surface area of the curve is also a measure of the AUC (the area under the ROC curve). Their indices are summarised in Table 2.7.

The results of analysis suggest that discriminant performance is higher for blue light stimuli, in particular for the b100 condition. The ROC curves show step-

Fig. 2.10 Comparison of ROC curves

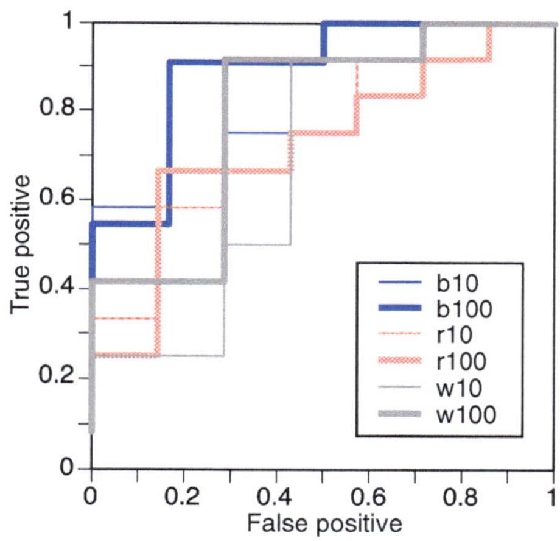

Table 2.7 Performance of models

Stimulus	AIC	R^2	Accuracy	AUC
b10	25.16	0.26	82.1	0.82
b100	17.24	0.47	89.4	0.89
r10	27.33	0.17	73.8	0.74
r100	28.05	0.14	72.6	0.73
w10	27.35	0.18	69.0	0.69
w100	24.70	0.28	79.8	0.80
b(10+100)	21.07	0.48	89.4	0.89
r(10+100)	30.26	0.22	71.4	0.71
w(10+100)	22.94	0.47	90.5	0.91
b+r	18.50	0.73	100	1.00
r+w	29.70	0.50	95.2	0.95
b+w	18.00	0.73	100	1.00
b+r+w	26.00	0.73	100	1.00

wise changes, since the number of participants influenced the results. However, the performance of AUC for b100 produced the highest reaction of any single stimulus condition.

2.6.2.2 Model Parameter Estimation

Table 2.4 shows the possibility of discriminating between AD patients and control group subjects using a logistic regression function. However, the parameters of these functions cannot be estimated sufficiently because the amount of data is too limited. In order to obtain accurate parameters for the estimation function, a Bayesian technique such as the Markov chain Monte Carlo (MCMC) method was introduced. This technique generates distributions of parameters, and thus the most appropriate parameters can be extracted. As a result, a more accurate estimation of the parameters can be conducted. In regard to the data generation procedure using the MCMC technique, the burn-in period was 2000, and the number of samples was 10000 [4]. The parameters were estimated from the distributions generated in the 8000 samples which remained after convergences were confirmed. Using this procedure for the b100 condition, the parameters are estimated as follows:

$$\hat{y}_i = 5.4229 + 0.4722 * factor1_i + 7.3801 * factor2_i$$

The magnitude of the coefficient for the second factor ($factor$ 2) is 15 times that of the coefficient for the first factor ($factor$ 1).

As an additional example, the parameters for blue light stimuli were estimated as follows:

$$\begin{aligned}\hat{y}_i = 9.1664 + 2.5320 * factor1_{i,b10} - 1.1366 * factor2_{i,b10} \\ -1.4011 * factor1_{i,b100} + 12.0891 * factor2_{i,b100}\end{aligned}$$

The magnitude of the parameters suggests that the coefficient for the second factor at a high level of brightness (b100) is relatively larger than for others (b10 and the first factor at a low level of brightness). As the patterns of coefficients depend on the light stimulus, the wavelengths of the stimuli may affect these reactions.

In regard to the discussion in the previous section, the probability of correct discrimination between control group subjects (1) or AD patients (0) may be illustrated using two-dimensional information ($factor$ 1 and $factor$ 2), as shown in Fig. 2.11. Figure 2.11 shows that the probability distribution against patients with AD depends mainly on the scores of the second factor. Also, the score of the first factor helps to more finely adjust the probability during the period where the curve is steep.

In regard to the experimental procedure, the features of PLR can be best measured during a 1 second high brightness level pulse of blue light (b100). Also, factor scores were calculated using the factor loading matrix shown in Table 2.6.

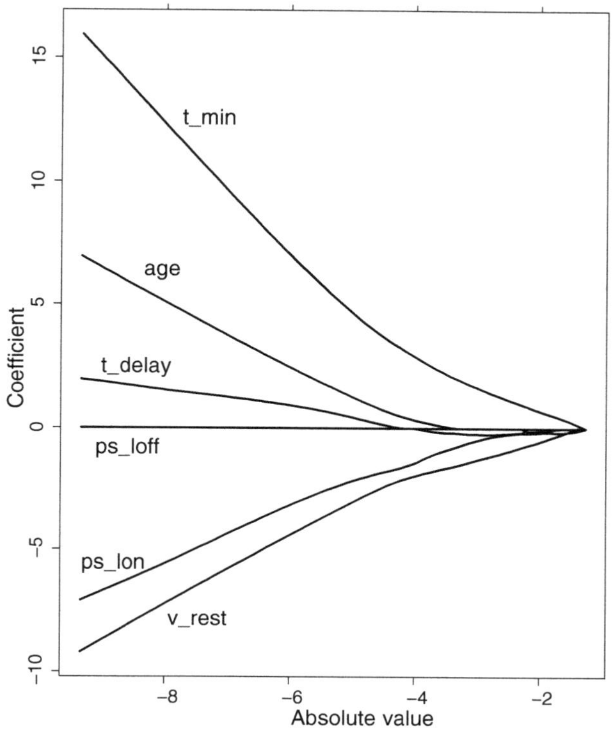

Fig. 2.11 Probability distribution for b100

Finally, the probability of diagnosing AD patients can be predicted using the above function.

However, the possibility of developing a more flexible procedure for use in future experiments involving additional new participants will be a subject of our further study.

2.6.3 LASSO Technique Application

2.6.3.1 Optimisation Using the LASSO Technique

In order to improve prediction performance while the sparse distribution of pupillary features for PLRs is considered, an optimisation procedure such as LASSO (Least Absolute Shrinkage and Selection Operator) [5, 7, 36] is introduced for discriminating AD patients and controlled subjects. In the previous section, factor analysis was introduced to reduce the dimensions of the features of PLRs. Two factor scores cover 63∼84% of the overall features, as shown in Table 2.6. In this section, the

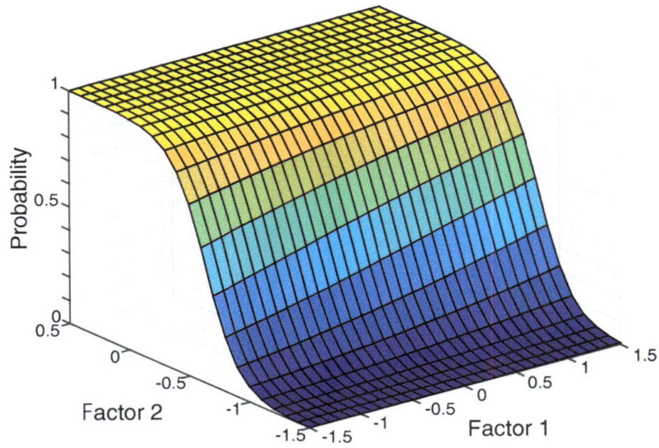

Fig. 2.12 Change in λ coefficients (b10 condition)

features are selected, and their performance is examined. Discrimination is made using a logistic regression function with the features of PLRs, as was done in the previous section. Since a logistic function is introduced as a link function, the probability of AD patients can be calculated. Here, performance is evaluated as a binary classification in order to optimise feature selection from all of the features as was done in the previous section. The LASSO technique calculates that some of the coefficients are driven to zero, thus making feature selection easy using the algorithm. The optimisation of selected variables and their weights for the prediction function using logistic regression is calculated using the LASSO procedure [18, 28] with a package "glmnet" [1]. An example of optimisation progress for a condition with a stimulus 10 cd/m^2 blue light (b10) is illustrated in Fig. 2.12. The horizontal axis indicates parameter λ, and the vertical axis indicates weights of variables (β). As the figure shows, weights for all variables converge to zero along with the parameter λ. This indicates a variable selection procedure. When a certain value of λ is optimised, the appropriate weights for the variables are extracted from the features of PLRs.

2.6.3.2 Results of the Optimisation with LASSO

Optimisation was conducted using AIC values which can be calculated using an index "sAIC" [3]. The optimised functions for each chromatic stimulus and for combinations of conditions, such as a blue light condition consisting two light intensities of 10 cd/m^2 and 100 cd/m^2, were calculated. The optimised weights (βs) at the minimised AIC are summarised in Table 2.8. Since the minimum parameter λ is produced by calculating the tenfold cross-validation (CV) value, the optimised weights are influenced by variances during the procedure. Another set of AICs and CV values at the minimum λ is attached to the table.

Table 2.8 Coefficients of β for features of PLR waveforms

Features	Blue		Red		White		Blue	Red	White
variables	b10(N=19)	b100(N=17)	r10(N=18)	r100(N=19)	w10(N=19)	w100(N=19)	b10&b100	r10&r100	w10&w100
ps_base	–	–	–	–	–	–	-/-	-/-	-/-
ps_lon	– 5.20	–	– 2.74	–	–	–	-/-	-/-	-/-
ps_loff	–	–	–	–	– 6.01	–	-/-	-/-	-/-
ps_min	–	– 6.81	– 1.75	– 10.81	–	– 1.57	-/-	-/ – 11.16	-/ – 2.36
RA	–	–	–	–	–	–	-/-	-/-	-/-
v_con	–	–	–	– 0.50	–	–	-/2.57	-/-	-/-
v_rest	– 6.73	–	–	– 5.23	–	–	– 1.31/-	-/ – 0.88	-/-
ac_max	–	8.64	11.20	–	1.10	15.05	-/4.27	-/-	-/5.56
t_delay	1.43	– 5.63	– 4.21	– 1.15	–	– 10.59	-/ – 1.85	– 2.37/-	-/ – 1.14
t_min	11.43	3.80	–	3.54	0.82	– 4.30	4.53/-	0.87/2.08	-/2.95
t_v_con	–	–	– 3.13	– 2.74	– 3.67	1.75	-/-	-/-	– 3.44/-
t_v_rest	–	0.34	3.35	2.67	8.45	2.02	-/-	-/-	0.53/0.85
int_con	–	–	–	–	–	–	-/-	-/-	-/-
int_rest	–	0.35	–	–	–	1.76	-/-	1.58/-	-/-
age	4.64	– 0.28	– 2.64	– 1.46	–	–	(-)	(-)	(-)
Intercept	– 2.29	– 5.88	– 4.79	– 4.16	– 3.38	– 5.40	– 2.84	– 4.22	– 2.94
AIC[#]	0.11	0.25	1.41	1.17	0.60	0.02	0.48	0.76	0.48
AUC	1.00	0.92	0.83	0.92	0.92	1.00	1.00	0.93	0.83
AIC[*]	4.13	10.56	15.19	10.46	11.36	5.14	5.05	12.33	8.53
CV error[*]	0.12	0.10	0.19	0.06	0.06	0.05	0.01	0.07	0.08

#: minimum of AIC
*: AIC and error rates in cross-validation method using the minimum λ as an optimum value
(-): "age" was not introduced for converged conditions

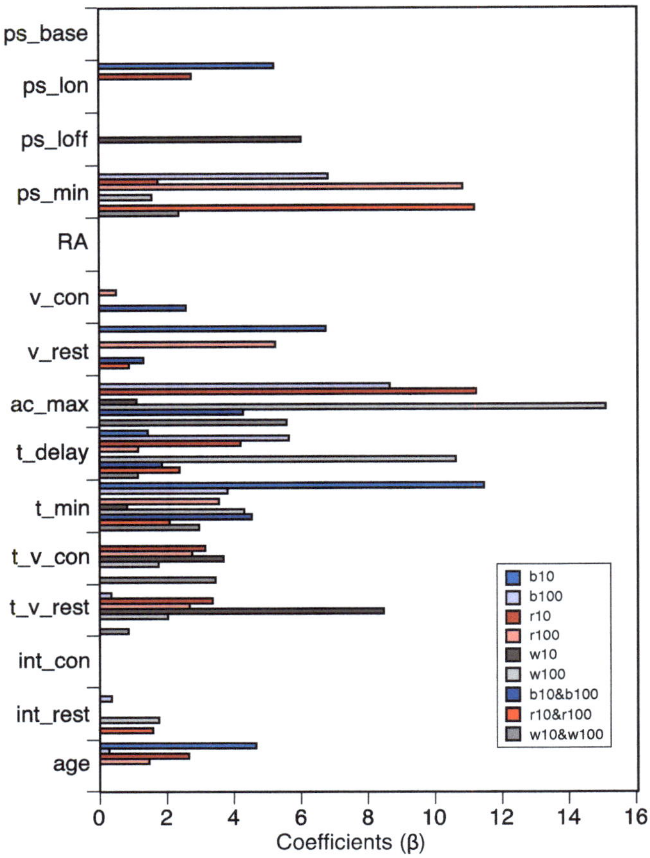

Fig. 2.13 Comparison of coefficients of features using prediction functions

The selected variables and their weights are summarised in Fig. 2.13. The bars indicate the weights of the nine conditions, while the variables that have been omitted have no weights. In comparing the number of variables, more variables were selected for conditions of 100 cd/m^2 than for the condition of 10 cd/m^2. Some of the selected variables concern factors for time and speed of pupillary change. Also, the ages of participants for blue and red light conditions were selected when the variable "age" is added to the selectable variables.

In evaluating AIC as a model of fitness for single light conditions (b10, b100, r10, r100, w10 and w100), the minimum AIC occurs under conditions "w100" and "b10", as a result of optimisation using AIC. The area under the curve (AUC) is 1.00 for both conditions when a different method of classification performance is used. The CV rates are zero for all conditions using this procedure. When the optimum solution is based on the minimum of λ, the condition for "b10" is the lowest of all

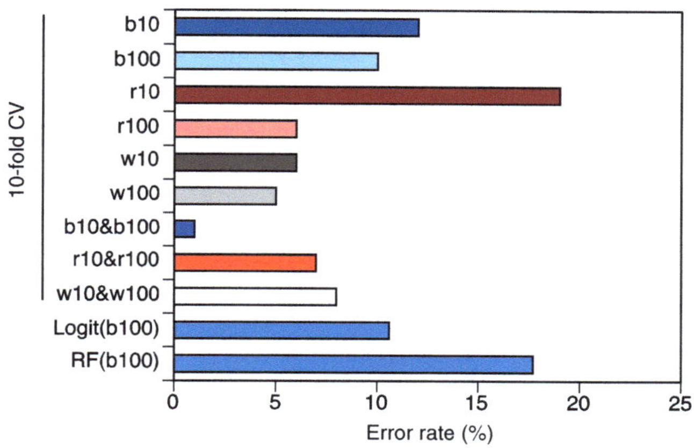

Fig. 2.14 Comparison of error rates for detecting AD patients (logit: previous section, RF: performance using Random Forest [42])

AICs for all of the single light conditions. The CV rates for "w10" and "w100" are the lowest for any of the conditions.

The evaluation was extended to converged conditions that converged (b10&b100, r10&r100 and w10&w100). The CV rates mostly decrease though the AICs increase with the number of variables, however. The tenfold CV rates are summarised in Fig. 2.14. The error rates using logistic function with factor scores from the previous section and using the random forest technique [42] are also displayed at the bottom of the bar graphs. As the figure shows, the error rate for the converged condition "b10&b100" is the lowest, and this rate is significantly lower than the prediction performance using logistic regression with factor scores and Random Forest technique.

2.6.4 Discussion

When logistic regression analysis was introduced, the probability of classifying AD patients was calculated using features extracted from PLRs. Prediction accuracy improved to nearly perfect using generated factor scores or some specific features of PLRs. The results confirm that some variables concerning factors for time and speed of pupillary changes contribute to the prediction of the probability of diagnosing patients with AD. In comparing the prediction performance with the performance in previous studies, the accuracy reaches to the levels of comparable without considering the sample size. However, as performance is based on the surveyed data set in this work, a more generalised procedure may be needed using larger clinical samples. Before gathering increased numbers of samples, experimental conditions for stimulus using different colours of light and at varying intensities and the validity of various features should be examined. These points are discussed as follows.

As shown at the end of the previous section, the prediction performance of the proposed procedures in our study should be compared and discussed. In some cases, performance using the LASSO technique is higher than for other techniques. When LASSO is applied to the converged condition "b10&b100", the error rate for tenfold CV values can be reduced to 0.01. Also, some of the conditions can produce error rates lower than 0.10, as shown in Table 2.8. The results suggest that an optimisation technique can reduce the error rate of prediction performance for AD patients using PLR features. Also, optimisation of prediction using the LASSO technique differs from other techniques. Conventionally, the contribution of the colour of the light, in particular blue light, is key to accurate prediction. Performance using PLR features of blue lights is the highest, while performance using features of white lights with conventional techniques is not high at all. As white light covers a wide range of light wavelengths, which stimulate ganglion cells on the retina, it is thought that the specific responses may not be detected. When the LASSO technique is introduced, error rates can also be suppressed, as is shown in Fig. 2.14. Therefore, the technique can select the significant variables that are capable of predicting AD patients.

In order to confirm the weight selectivity that was hypothesised, weight patterns of three procedures for variables with the "b100" condition are compared in Table 2.9. Here, weight patterns using LASSO, importance values using Random Forest and factor loadings using Bayes technique with factor scores are compared. The last factor loadings depend on whole responses, so they are not specific values with the "b100" condition.

Table 2.9 Comparison of weights of variables (b100 condition)

	LASSO logistic regression	Random forest importance	Factor+Bayes F1/F2
ps_base	–	0.37	0.60 / 0.59
ps_lon	–	0.28	0.60 / 0.59
ps_loff	–	0.19	– 0.08 / 1.00
ps_min	– 6.81	0.84	– 0.11 / 1.02
RA	–	0.19	1.00 / – 0.06
v_con	–	0.81	– 0.91 / – 0.07
v_rest	–	1.24	0.64 / 0.10
ac_max	8.64	0.84	– 0.90 / – 0.02
t_delay	– 5.63	0.09	– 0.71 / 0.06
t_min	3.80	0.80	0.09 / – 0.64
t_v_con	–	0.28	– 0.74 / 0.11
t_v_rest	0.34	0.48	0.07 / – 0.44
int_con	–	0.61	0.99 / 0.01
int_rest	0.35	0.13	1.01 / – 0.27
Error rate	0	17.7	10.6

As Table 2.9 shows, selected variables using LASSO have certain weights using the two other techniques, but some of the non-selected variables also have large weights. Regarding prediction performance using error rates, the LASSO technique may be able to select significant variables, as its performance is the highest. It may be a benefit of the technique, as the effectiveness of the LASSO technique has been confirmed. However, a more detailed analysis of the means to improve performance is needed to establish a more robust procedure for predicting outcomes in AD patients, which will be the subject of our further study.

2.7 Conclusion

This chapter confirms the possibility of detecting AD patients using a set of features of PLRs in response to chromatic stimuli. In addition, procedures for predicting the probability of diagnosing AD patients are developed using features of PLRs, which respond to the activity of ipRGCs. Three colours of light at two levels of brightness were illuminated for 1 second, and pupil light responses were observed and recorded. Fifteen features were extracted from each PLR, and two factor scores were calculated using a factor loading matrix. These features were extracted and measured using the Fourier descriptor technique for PLR waveform shapes, and specific features of PLRs were also defined. In order to improve the performances, two techniques were introduced: logistic regression and LASSO technique for optimisation. The following results were obtained.

1. The differences in features of PLRs between the two groups of participant were compared. In the results, there were no differences in Fourier descriptors, but some significant differences in features of PLRs were produced for the b100 and the r10 experimental conditions.
2. PLRs for factors related to ageing were compared, and there were some significant differences in features of PLRs in regard to response delay and range of pupillary change. These differences were observed mainly when a white light stimulus was used.
3. Prediction analysis for AD patients was conducted using the Random Forest technique, and most participants in the healthy control group were estimated correctly. The estimation performance was the highest when the features of PLRs for blue lights were employed. This confirms the contribution of ipRGCs to the prediction of AD patients.
4. The variables that contributed to the highest levels of correct estimation were identified. They represented the temporal behaviours of PLRs.
5. There are significant differences in some features between AD patients and control group subjects, in particular for the b100 condition. Also, for a few features for white light there are significant differences between age levels.
6. Logistic regression analysis was introduced to discriminate AD patients from the control group using two factor scores in response to chromatic stimuli.

Performance was evaluated using the indices of the fitness of equations. In the results, the performance for b100 was the highest.
7. The MCMC technique was introduced to estimate the parameters of the regression functions. The model provides a distribution of probability for AD patients and the control group.
8. In order to predict AD patients using the extracted features of PLRs, LASSO technique was introduced to logistic regression analysis. As a result, the error rate for prediction can be reduced to 0.01 using a converged condition of "b10&b100" where significant variables of features are selected.
9. In comparing weight patterns for feature variables of PLRs, the features that contributed to prediction were discussed.

The validity of the probability estimations should be confirmed using the PLR data of existing and additional new patients. This will be a subject of our further study.

Acknowledgments This research was partially supported by Polish Ministry of Science and Higher Education research grant NN518 405338.

This work is an extended version that is based on reports given at FedCSIS2018 [41]. The authors would like to thank those who provided useful comments regarding the conference paper.

References

1. Package 'glmnet'. https://cloud.r-project.org/web/packages/glmnet/glmnet.pdf
2. Package: randomForest. http://stat-www.berkeley.edu/users/breiman/RandomForests
3. Package 'saic'. https://sites.google.com/site/shuichikawanoen/
4. SAS/STAT 13.1 user's guide, the MCMC procedure. http://support.sas.com/documentation/onlinedoc/stat/131/mcmc.pdf
5. Akaho, S.: Kernel Multivariate Analysis: New Trends in Nonlinear Data Analysis. IWANAMI SHOTEN, Tokyo (2008)
6. Aoki, K., Ngo, T.T., Mitsugami, I., Okura, F., Niwa, M., Makihara, Y., Yagi, Y., Kazui, H.: Early detection of lower MMSE scores in elderly based on dual-task gait. IEEE Access **7**, 40085–40094 (2019)
7. Bishop, C.M.: Pattern Recognition and Machine Learning. Springer Science+Business Media, LLC, New York (2006)
8. Bitsios, P., Prettyman, R., Szabadi, E.: Changes in autonomic function with age: a study of pupillary kinetics in healthy young and old people. Age Ageing **25**, 432–438 (1996)
9. Bittner, D.M., Wieseler, I., Wilhelm, H., Riepe, M.W., Müller, N.G.: Repetitive pupil light reflex: Potential marker in Alzheimer's disease? J. Alzheimer's Dis. **42**, 1469–1477 (2014)
10. Chougule, P.S., Najjar, R.P., Finkelstein, M.T., Kandiah, N., Milea, D.: Light-induced pupillary responses in Alzheimer's disease. Front. Neurol. **10**(360), 1–12 (2019)
11. Dacey, D.M., Liao, H.W., Peterson, B.B., Robinson, F.R., Smith, V.C., Pokorny, J., Yau, K.W., Gamlin, P.D.: Melanopsin-expressing ganglion cells in primate retina signal color and irradiance and project to the LGN. Nature **433**, 749–754 (2005)
12. Ding, J.D., Lin, J., Mace, B., Herrmann, R., Sullivin, P., Rickman, C.: Targeting age-related macular degeneration with Alzheimer's disease based immunotherapies: Anti-amyloid-β antibody attenuates pathologies in an age-related macular degeneration mouse model. Vis. Res. **48**, 339–345 (2008)

13. Feigl, B., Mattes, D., Thomas, R., Zele, A.J.: Intrinsically photosensitive (melanopsin) retinal ganglion cell function in glaucoma. Invest. Ophthalmol. Vis. Sci. **52**(7), 4362–4367 (2011)
14. Feigl, B., Zele, A.J.: Melanopsin-expressing intrinsically photosensitive retinal ganglion cells in retinal disease. Ophthalmol. Vis. Sci. **91**(8), 894–903 (2014)
15. Folstein, M.F., Folstein, S.E., McHugh, P.R.: MINI-MENTAL STATE—a practical method for grading the cognitive state of patients for the clinician. J. Psychiatr. Res. **12**, 189–198 (1975)
16. Fotiou, D.F., Setergiou, V., Tsiptsios, D., Lithari, C., Nakou, M., Karlovasitou, A.: Cholinergic deficiency in Alzheimer's and Parkinson's disease: evaluation with pupillometry. Int. J. Psychophysiol. **73**, 143–149 (2009)
17. Fotiou, F., Fountoulakis, K., Tsolaki, M., Goulas, A., Palikaras, A.: Changes in pupil reaction to light in Alzheimer's disease patients: a preliminary report. Int. J. Psychophysiol. **37**, 111–120 (2000)
18. Friedman, J., Hastie, T., Tibshirani, R.: Regularization paths for generalized linear models via coordinate descent. J. Stat. Softw. **33**, 1–22 (2010)
19. Frost, S., Kanagasingam, Y., Sohrabi, H., Bourgeat, P., Villemagne, V., Rowe, C.C., Macaulay, L.S., Szoeke, C., Ellis, K.A., Ames, D., Masters, C.L., Rainey-Smith, S., Martins, R.N., Group, A.R.: Pupil response biomarkers for early detection and monitoring of Alzheimer's disease. Curr. Alzheimer Res. **10**(9), 931–939 (2013)
20. Gamlin, P.D., McDougal, D.H., Pokorny, J.: Human and macaque pupil responses driven by melanopisn-containing retinal ganglion cells. Vis. Res. **47**, 946–954 (2007)
21. Granholm, E.L., Panizzon, M.S., Elman, J.A., Jak, A.J., Hauger, R.L., Bondi, M.W., Lyons, M.J., Franz, C.E., Kremen, W.S.: Pupillary responses as a biomarker of early risk for Alzheimer's disease. J. Alzheimer's Dis. **56**(4), 1419–1428 (2017)
22. Hatori, M., Panda, S.: The emerging roles of melanopsin in behavioral adaptation to light. Trends Mol. Med. **16**(10), 435–446 (2010)
23. Hattar, S., Liao, H.W., Takao, M.: Melanopsin-containing retinal ganglion cells: architecture, projections, and intrinsic photosensitivity. Science **295**, 1065–1070 (2002)
24. Heneka, M.T., Nadrigny, F., Regen, T., Martinez-Hernandez, A., Dumitrescu-Ozimek, L., Terwel, D., Jardanhazi-Kurutz, D., Walter, J., Kirchhoff, F., Hanisch, U.K., Kummer, M.P.: Locus coeruleus controls Alzheimer's disease pathology by modulating microglial functions through norepinephrine. PNAS **107**(13), 6058–6063 (2010)
25. Iijima, A., Haida, M., Ishikawa, N., Ueno, A., Minamitani, H., Shinohara, Y.: Re-evaluation of tropicamide in the pupillary response test for Alzheimer's disease. Neurobiol. Aging **24**, 789–796 (2003)
26. Jarrold, W., Peintner, B., wilkins, D., Vergryi, D., Richey, C., Gorno-Tempini, M.L., Ogar, J.: Aided diagnosis of dementia type through computer-based analysis of spontaneous speech. In: Proceedings of Workshop on Computational Linguistics and Clinical Psychology: From Linguistic Signal to Clinical Reality, pp. 27–37. Association for Computational Linguistics, New York (2014)
27. Kardon, R.H., Anderson, S.C., Damarjian, T.G., Grace, E.M., Stone, E., Kawasaki, A.: Chromatic pupillometry in patients with retinitis pigmentosa. Ophthalmology **118**(2), 376–381 (2011)
28. Kawano, S., Matsui, H., Hirose, K.: Statistical Modeling via Sparse Estimation. KYORITSU Publication, Tokyo (2018)
29. Kawasaki, A., Kardon, R.H.: Intrinsically photosensitive retinal ganglion cells. J. Neuroophthalmol. **27**, 195–204 (2007)
30. Klumpp, P., Fritsch, J., Noeth, E.: ANN-based Alzheimer's disease classification from bag of words. In: Proceedings of Speech Communication; 13th ITG-Symposium, pp. 1–4. IEEE, New York (2018)
31. Lim, J.K.H., Li, Q.X., He, Z., Vingrys, A.J., Wong, V.H., Currier, N., Mullen, J., Bul, B.V., Nguyen, C.T.O.: The eye as a biomarker for Alzheimer's disease. Front. Neurol. **10**(536), 1–14 (2016)

32. Lucas, R.J., Peirson, S.N., Berson, D.M., Brown, T.M., Cooper, H.M., Czeisler, C.A., Figueiro, M.G., Gamlin, P.D., Lockley, S.W., O'Hagan, J.B., Price, L.L., Provencio, I., Skene, D.J., Brainard, G.C.: Measuring and using light in the melanopsin age. Trends Neurosci. **37**(1), 1–9 (2014)
33. Morgia, C.L., Ross-Cisneros, F.N., Hannibal, J., Montagna, P., Sadun, A.A.: Melanopsin-expressing retinal ganglion cells: implications for human diseases. Vis. Res. **51**, 296–302 (2011)
34. Morgia, C.L., Ross-Cisneros, F.N., Koronyo, Y., Hannaibal, J., Gallassi, R., Cantalupo, G., SAmbati, L., Pan, B.X., Tozer, K.R., Barboni, P., Provini, F., Avanzini, P., Carbonelli, M., Pelosi, A., Chui, H., Liguori, R., Baruzzi, A., Koronyo-Hamaoui, M., Sadun, A.A., Carrelli, V.: Melanopsin retinal ganglion cell loss in Alzheimer disease. Ann. Neurol. **79**(1), 90–109 (2016)
35. Morishita, I., Kobatake, H.: Signal Processing (In Japanese). The Society of Instrument and Control Engineers, Tokyo (1982)
36. Murphy, K.P.: Machine Learning–A Probabilistic Perspective. The MIT Press, Massachusetts (2012)
37. Murphy, P.R., III H LeVine: Alzheimer's disease and the β-amyloid peptide m. J. Alzheimer's Dis. **19**, 311 (2010)
38. Murphy, P.R., Robertson, I.H., Balsters, J.H., O'Connell, R.G.: Pupillometry and P3 index the locus coeruleus-noradrenergic arousal function in humans. Psychophysiology **48**(11), 1532–1543 (2011)
39. Nakayama, M., Nowak, W., Ishikawa, H., Asakawa, K., Ichibe, Y.: Discovering irregular pupil light responses to chromatic stimuli using waveform shapes of pupillograms. EURASIP J. in Bioinformatics and System Biology **#18**, 1–14 (2014)
40. Nowak, W., Żarowska, A., Szul-Pietrzak, E., Misiuk-Hojło, M.: System and measurement method for binocular pupillometry to study pupil size variability. BioMedical Eng. Online **13**(#69), 1–16 (2014)
41. Nowak, W., Nakayama, M., Kręcicki, T., Hachoł, A.: Prediction of Alzheimer disease in patients using features of pupil light reflex to chromatic stimuli. In: Proceedings of the Federated Conference on Computer Science and Information Systems (FedCSIS), pp. 269–273 (2018)
42. Nowak, W., Nakayama, M., Kręcicki, T., Trypka, E., Andrzejak, A., Hachoł, A.: Analysis for extracted features of pupil light reflex to chromatic stimuli in Alzheimer's patients. EAI Endorsed Trans. Pervasive Health Technol. **5**, 1–10 (2019). E4
43. Nowak, W., Nakayama, M., Pieniążek, M., Hachoł, A.: Feature analyses of pupil light reflex to chromatic stimuli in Alzheimer's patients. In: Proceedings of 2nd International Conference on Frontiers of Signal Processing, pp. 58–62 (2016)
44. Ohno-Matsui, K.: Parallel findings in age-related macular degeneration and Alzheimer's disease. Prog. Retin. Eye Res. **30**, 217–238 (2011)
45. Park, J.C., Moura, A.L., Raza, A.S., Rhee, D.W., Kardon, R.H., Hood, D.C.: Toward a clinical protocol for assessing rod, cone, and melanopsin contributions to the human pupil response. Invest. Ophthalmol. Vis. Sci. **52**(9), 6624–6635 (2011)
46. Pinkowski, B.: Robust Fourier descriptions for characterizing amplitude-modulated waveform shapes. J. Acoust. Soc. Am. **95**(6), 3419–3423 (1994)
47. Prettyman, R., Bitsios, P., Szabadi, E.: Altered pupillary size and darkness and light reflexes in Alzheimer's disease. J. Neurol. Neurosurg. Psychiatry **62**, 665–668 (1997)
48. Prince, M., Jackson, J.: World Alzheimer Report 2009. Alzheimer's Disease International, London (2009). https://www.alz.co.uk/research/files/WorldAlzheimerReport.pdf
49. Satoh, A., Iijima, K.M.: Roles of tau pathology in the locus coeruleus (LC) in age-associated pathophysiology and Alzheimer's disease pathogenesis: potential strategies to protect the LC against aging. Brain Research (2017). https://doi.org/10.1016/j.brainres.2017.12.027. http://www.sciencedirect.com/science/article/pii/S0006899317305620
50. Sivak, J.M.: The aging eye: Common degenerative mechanisms between the Alzheimer's brain and retinal disease. Invest. Ophthalmol. Vis. Sci. **54**(1), 871–880 (2013)

51. Steinerman, J., Irizarry, M., Scarmeas, N., Raju, S., Brandt, J., Albert, M., Blacker, D., Hyman, B., Stern, Y.: Distinct pools of beta-amyloid in Alzheimer disease-affected brain: a clinicopathologic study. Arch Neurol. **65**, 906–912 (2008)
52. Tombaugh, T., McDowell, I., Kristjansson, B., Hubley, A.: Mini-mental state examination (MMSE) and the modified MMSE (3MS): a psychometric comparison and normative data. Psychol. Assess. **8**(1), 48–59 (1996)
53. Villemagne, V.L., Burnham, S., Bourgeat, P., Brown, B., Ellis, K.A., Salvado, O., Szoeke, C., Macaulay, S.L., Martins, R., Maruff, P., Ames, D., Rowe, C.C., Masters, C.L., Group, A.R.: Amyloid β deposition, neurodegeneration, and cognitive decline in sporadic Alzheimer's disease: a prospective cohort study. Lancet Neurol. **12**, 357–367 (2013)
54. van Wijngaarden Franzco, P., Hadoux, X., Alwan, M., Keel, S., Dirani, M.: Emerging ocular biomarkers of Alzheimer disease. Clin. Exp. Ophthalmol. **45**(1), 54–61 (2017)
55. Yoshida, T., Ohno-Matsui, K., Ichinose, S., Sato, T., Iwasa, N., Saido, T.C., Hisatomi, T., Mochizuki, M., Morita, I.: The potential role of amyloid β in the pathogenesis of age-related macular degeneration. J. Clin. Invest. **115**(10), 2793–2800 (2005)
56. Young, R., Kimura, E.: Pupillary correlates of light-evoked melanopsin activity in humans. Vis. Res. **48**, 862–871 (2008)
57. Zele, A.J., Adhikari, P., Cao, D., Feigl, B.: Melanopsin and cone photoreceptor inputs to the afferent pupil light response. Front. Neurol. **10**(529), 1–9 (2019)
58. Zhang, D., Lu, G.: A comparative study on shape retrieval using Fourier descriptors with different shape signatures. In: Proceedings of the 5th Asian Conference on Computer Vision, pp. 646–651. Springer, Berlin (2002)
59. Zhou, L., Fraser, K.C., Rudzicz, F.: Speech recognition in Alzheimer's disease and in its assessment. In: Proceedings of INTERSPEECH 2016, pp. 1948–1952. ISCA, New York (2016)

Chapter 3
Classification of Participants Using Metrics of Oculo-motors During Dementia Tests

Abstract Ocular information was observed during a set of dementia tests involving participants with Alzheimer's disease (AD), with a mild level of cognitive impairment (MCI) or in a control group. The number of participants was 26. Features of changes in pupil size and in the central position of both eyes of participants of all three types were compared. There are significant differences in some of the metrics between the types, in earlier test sessions. The possibility of classification was confirmed using the extracted features, and the contributions of some features were examined. During the Mini-Mental State Examination (MMSE) surveillance, 11-dimensional ocular information was measured in six tests. The 26 elder participants were diagnosed as Alzheimer's disease (AD), mild cognitive impairment (MCI) and normal control (NC). The extracted features were compared between participant's groups and tests. In order to predict participant's group using the feature values and random forest technique, the training data set was generated using the statistic in each group. Feature selections were conducted to optimise the prediction performance. As the results, ordinal eye movement features were selected to estimate participant's groups.

Chapter Co-authors: Wioletta Nowak, Minoru Nakayama, Trypka, Elżbieta, and Żarowska, Anna

Originally published as Wioletta Nowak, Minoru Nakayama, Trypka, Elżbieta, Żarowska, Anna, Classification of Alzheimer's Disease Patients using Metrics of Oculo-Motors, Proceedings of the 16th Conference on Computer Science and Intelligent Systems, pp. 403–407, IEEE, 2021.
Wioletta Nowak, Minoru Nakayama, Trypka, Elżbieta, Żarowska, Anna, An estimation of disease levels using ocular information during Dementia tests, The Transactions of Human Interface Society, Vol. 26, Issue 3, pp.301–304, 2024.

M. Nakayama, W. Nowak, *Development of Diagnostic Procedures Using Features of Pupil Reactions*, Behaviormetrics: Quantitative Approaches to Human Behavior 21, https://doi.org/10.1007/978-981-95-4502-5_3

3.1 Introduction

Human cognitive ability is often affected by disease, such as by Alzheimer's disease (AD) for example. The decline in cognitive ability of elderly people can be monitored carefully during daily exercise session [1]. In general, behavioural actions requiring some level of workload affect visual information processing and cognitive perception. The symptoms of diseases can be detected using biometrics, such as indices of oculo-motors. Changes in the level of cognitive activity and mental workload handling ability are often assessed using pupillary changes [2, 5, 7]. The symptoms of AD and aged macular disease (AMD) are also detected using pupillary reactions to the pupil light reflex (PLR) [8, 10, 11].

Most diagnostics for AD patients are based on medical consultations using cognitive tests such as the Mini-Mental State Examination (MMSE) [4] or the Montreal Cognitive Assessment (MoCA) [9]. In the test results, participants are classified into AD, mild cognitive impairment (MCI) and no cognitive impairment (NCI) or categories with normal elderly controls (NCs) [9]. Since medical consultations are conducted during face-to-face oral questioning, some communication skills such as language ability and the participant's mental situation may influence the diagnostic result. Therefore, the time spent by medical practitioners and the workload of participants is not insignificant. Any appropriate procedures that can assist with medical observation using participant's behavioural metrics to reduce the difficulty of making proper assessments and permit this AD diagnostic procedure become more widely available.

As mentioned above, ocular responses including pupillary changes and eye movement provide some evidence of symptoms of cognitive activity impairment. Observation of participant's oculo-motor changes is introduced into medical consultation, and the possibility of detecting cognitive impairment is examined. The following topics are addressed in this chapter.

1. Pupillary changes and eye movement are measured using oculo-motor indices during a medical consultation to diagnose MMSE, and the metrics of the three diagnostic classes AD, MCI and NC are compared.
2. A prediction procedure for diagnostic classes is developed using the metrics measured, and its performance is discussed.

3.2 Experimental Method

During a medical consultation of elderly people for MMSE, ocular metrics were measured.

3.2.1 Cognitive Test

A cognitive exam, based on MMSE and consisting of the following six tests, was administered.

- Test 1: Control condition for measurement without any tasks or baseline monitoring
- Test 2: Orientation including simple questions and a conversation session
- Test 3: Memory and recall tests of five words
- Test 4: Oral calculation such as decrements of 7 starting from 100
- Test 5: Second question and conversation session
- Test 6: Recall test for words memorised in Test 3

The tests were conducted at a hospital in Poland, and each participant was examined for AD and MCI by an M.D. who was one of the authors. Therefore, test tasks and duration varied by participant according to their medical condition.

3.2.2 Participants

In the results of the diagnoses, 26 participants (mean age: 73.7 yo, STD: 9.9 yo) were classified into three groups. The normal control (NC) group consisted of people of comparable age.

- AD: 4 (F: 4, mean age: 77.3 yo, STD: 6.7 yo)
- MCI: 15 (F: 11, M: 4, mean age: 73.4 yo, STD 13.1 yo)
- NC: 7 (F: 5, M: 2, mean age: 72.4 yo, STD:5.3 yo)

Informed consent was obtained from all participants prior to the experiment. One MCI (F, 70 yo) had data partially missing, and this data was omitted from later analysis.

3.2.3 Procedure

Participants wore image tracker eye goggles (VisualEyes 505) and were asked to view the face of an M.D. during the test. Before the test, calibration of viewing points was conducted. Images of both eyes were recorded as 640×240 pixel images at 25 Hz using a camera with a lens in each side of the goggles. The images were recorded using the MP4 format. After the experiment, the centre of the pupil and the size of the pupil were extracted. Figure 3.1 shows the pupillary changes of an AD patient during Tests 1 through 6. The data measured during blinks was omitted in the analysis which followed. Durations where pupillary changes were dropped indicate data omitted due to blinks.

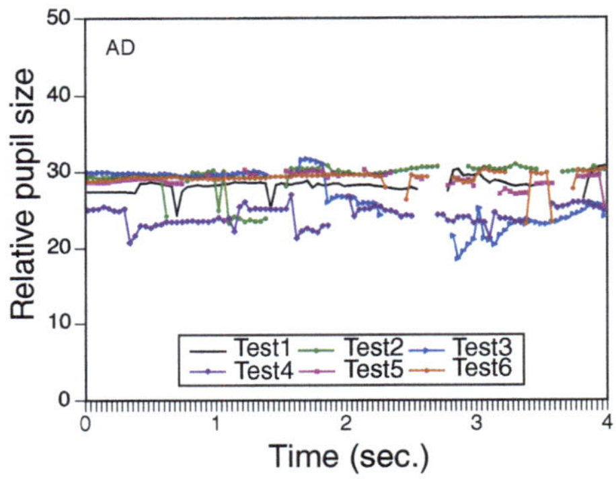

Fig. 3.1 Pupillary changes of an AD patient

3.2.4 Post-Processing

3.2.4.1 Pupil Response

Pupil size was calculated as the sum of the pixels of each frame of the black and white video using an image processing technique. Mean pupil sizes were calculated for each duration of observation. Pupillary changes were evaluated as frequency powers of less than 4 Hz, as pupils behave like low-pass filters. The frequency powers were calculated as mean power spectrum densities for 0.5∼1.6 Hz and 1.6∼4.0 Hz, as they were in the previous study [13]. The lowest frequency power (f <0.5 Hz) was omitted due to signal noise [15].

3.2.4.2 Eye Movement

During image processing, the centre of the pupil was measured. The measurement specification of the goggles suggests that the shift in the centre of the pupil corresponds with a visual angle of 0.25 deg for both the horizontal and vertical axes. Eye movement data was generated using this method of conversion. The eye movement was classified into fixations and saccades using a threshold of 40 deg/sec [3]. For each test observation, saccade frequency, mean saccade lengths, fixation frequency, and mean fixation duration were summarised.

Since the ocular metrics of both eyes were measured, they were evaluated using a procedure that examined repeated individual measurements.

3.3 Results of Features of Oculo-Motors and Prediction Performance

3.3.1 Observation Time

As the duration of test observation depended on the condition of the individual participant, mean durations are summarised in Table 3.1, which consists of test

Table 3.1 Mean observation duration (sec) for participant groups

	Test					
Subject group	1	2	3	4	5	6
AD (N=4)	44.6	58.4	65.7	61.1	57.1	76.4
(SD)	14.0	31.3	10.8	7.3	9.8	32.5
MCI (N=15)	33.4	62.7	53.1	39.6	36.0	42.8
(SD)	3.2	27.4	10.4	15.1	10.5	27.3
Control (N=7)	32.1	61.6	58.5	32.3	32.7	23.1
(SD)	1.1	17.5	24.4	9.5	6.3	9.1

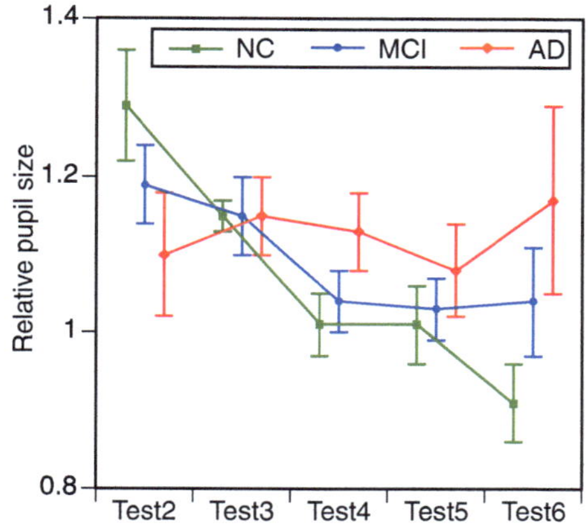

Fig. 3.2 Mean relative pupil sizes between groups (error bar: STD error)

number and participant group. The duration of most tests except for Test 2 was around 30 seconds. While the duration varied between tests, it increased gradually as group participants progressed from NC to MCI to AD. There may have been some difficulty for participants with AD to perform some of the test tasks.

3.3.2 *Pupillary Changes*

3.3.2.1 Mean Pupil Sizes

Pupil sizes were standardised in Test 1 using mean pupil sizes. Mean relative pupil sizes for the three groups of participants are summarised in Fig. 3.2. Error bars indicate standard errors. The mean size of NC decreased gradually from Tests 2 and 6. Both task difficulty and order effect may have affected pupil sizes. However, mean pupil size of the AD group remained relatively large throughout the tests, and the deviations were smaller than the deviations of other groups. The means for MCI also decreased along with the tests, and the deviation was smaller than for the NC group.

Two-way ANOVA was conducted in order to examine the contributions of the factors of participant groups and test results. These results are summarised in

Table 3.2 Two-way ANOVA: participant groups × Tests

Source	df	SS	MS	F	Sig.
Subject group	2	0.059	0.029	0.43	n.s.
Test	5	1.078	0.216	3.15	$p < 0.01$
Sub-G × Test	10	0.620	0.062	0.91	n.s.

Table 3.3 ANOVA analysis for each subject group

Subject group	df	SS	MS	F	Sig.
AD	5	0.154	0.031	0.35	n.s.
MCI	5	0.816	0.163	2.14	$p < 0.10$
NC	5	1.298	0.260	6.33	$p < 0.01$

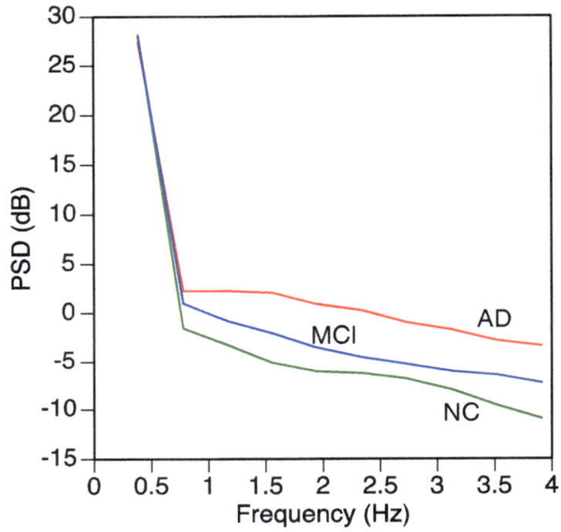

Fig. 3.3 PSD of pupillary change in Test 1

Table 3.2. The test factor is significant, but participant groups and the interaction of the two factors are not significant.

In further analysis, an F-test of the participants of each group was conducted. The results are summarised in Table 3.3. Test factors were significant for both NC ($p < 0.01$) and MCI ($p < 0.10$). The results may reflect the pupil responses of participants of the groups.

3.3.2.2 Pupillary Oscillation

As pupillary changes between tests and groups of participants were confirmed, frequency analysis was also applied. As mentioned in the section above, power spectrum densities (PSDs) in the frequency ranges from 0.5 to 1.6 Hz and from 1.6 to 4.0 Hz were compared. There were no significant changes between tests or groups of participants.

A typical change was observed during Test 1. The PSD of changes of each group is summarised in Fig. 3.3. As Fig. 3.3 shows, there are some differences between groups of participants. The results of F-tests of the means of PSDs are summarised

Table 3.4 ANOVA for participant groups in Test 1

Frequency	df	SS	MS	F	Sig.
0.5 < f < 1.6 Hz	2	6.887	3.444	2.62	$p < 0.10$
1.6 < f < 4 Hz	2	2.062	1.031	5.62	$p < 0.01$

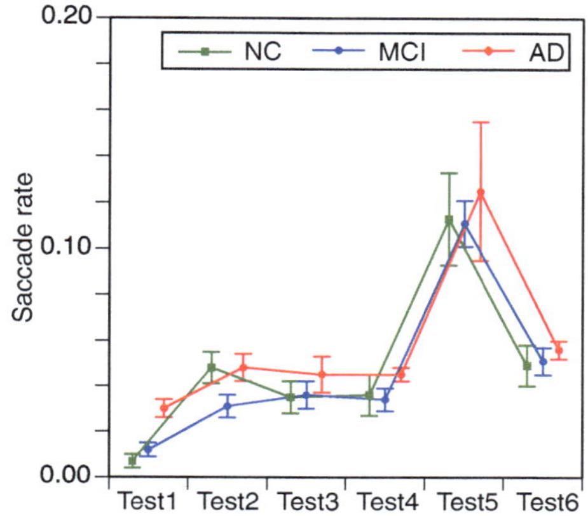

Fig. 3.4 Saccade frequency (deg/sec)

in Table 3.4. There is a significant difference ($p < 0.01$) between groups in the high frequency range (1.6∼4.0 Hz) and the tendency for a difference to exist ($p < 0.10$) in the low frequency range (0.5∼1.6 Hz).

This result suggests that differences in PSDs appear during controlled conditions, such as in Test 1.

3.3.3 Eye Movement

Features of eye movement, which were tracked using the above-mentioned procedure, are summarised for mean saccade frequency in Fig. 3.4. The error bars indicate STD errors.

In Fig. 3.5, there are few differences between participant groups, but mean frequencies increase gradually during medical consultation after Test 1. The change in frequencies during complete tests was examined using two-way ANOVA, and the factor for tests was significant ($F(5, 294) = 28.4$, $p < 0.01$); however the factor for groups of participants was not significant ($F(2, 294) = 1.8$, $p = 0.16$). The effect of groups of participants on each test was examined, while the factor in Test 1 ($F(2, 49) = 7.7$, $p < 0.01$) was significant, and there were significant differences between groups except for one pair of MCI and NC. Even in Test 2, the factor for groups of participants showed a tendency to be different ($F(2, 49) = 2.48$, $p < 0.10$), but overall there were no significant differences between groups.

Changes in mean saccade lengths are summarised in Fig. 3.5. The effect of groups of participants was only examined in Test 2 ($F(2, 48) = 3.0$, $p < 0.10$), but there were no significant differences between groups. Also, a significant different

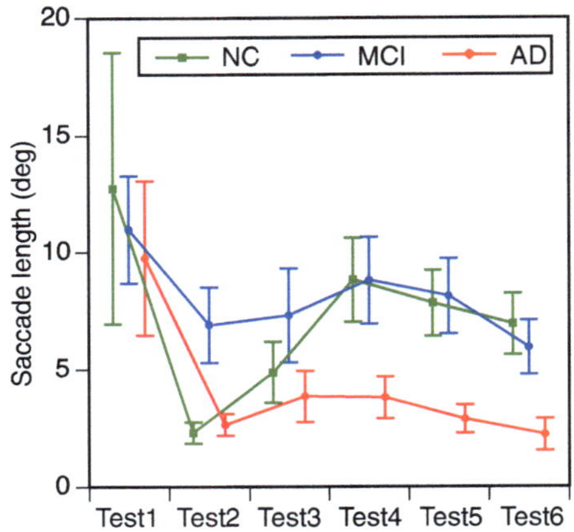

Fig. 3.5 Saccade length (deg)

tendency for groups of participants to be different was observed in the fixation frequency of Tests 1 and 2. Therefore, typical eye movement responses may appear in initial tests, as they did in Tests 1 and 2.

In this study, conversations during medical consultation were not analysed, though some mental workloads were detected in oculo-motor indices during Tests 2 through 6. The results suggest that there are significant differences between the metrics of Test 1 and the metrics of the other tests. However, the deviations in metrics measured during tests may not be large. Therefore, the metrics can be noted as means of duration for the tests.

3.3.4 *Prediction of Groups of Participants*

Since there were some significant differences in the metrics of the groups of participants extracted, the possibility of predicting the class of the group was examined using the features extracted.

3.3.4.1 Prediction Using the Observed Metrics

Three groups of participants were estimated in six tests using six metrics (mean pupil size, mean PSD of pupil oscillation, mean saccade frequency, mean saccade length, mean fixation frequency, and mean fixation duration). An estimation model was trained using the Random Forest method.

Estimation was conducted using the six-dimensional data from Tests 1 to 6, and performance was examined using trial and error. As a qualitative evaluation, performance during Tests 1 and 2 was better. The results are summarised in the contingency table shown in Table 3.5. The vertical cells represent groups

Table 3.5 Classification results using Tests 1 and 2

Subject group	(1)	(2)	(3)	Recall	F1	Prediction pairs
(1) AD	6	2	0	0.75	0.80	ad-ad:3/4
(2) MCI	0	25	3	0.89	0.85	mci-*:14/14
(3) NC	1	4	9	0.64	0.69	nc-*:5/7
Precision	0.86	0.81	0.75	P-mean:0.81; R-mean:0.76		

Rate of lower triangular matrix: 0.90

of participants, and the horizontal cells represent predictions. As the predictions were conducted using both eyes, the overall numbers are twice the number of participants. The prediction performance recall rate, precision rate and F1 indices were calculated. Also, the correct prediction by either eye is summarised in the rightmost cell of Table 3.5.

The results of the table suggest the possibility of predicting a participant's group using the metrics of eye observations. The purpose is a targeted diagnostic procedure which can predict the most-advanced levels of AD. Therefore, the number of predictions of the lower triangular matrix may be another index of performance. The highest recall rate appears with the MCI group, as the number of participants is the largest. The number of participants may be an influence bias regarding prediction performance.

3.3.4.2 Prediction with Data Extensions

The medical consultation data set seems insufficient because it is not easy to gather a large and balanced number of participants for each group. A data extension technique was introduced in order to obtain a set of prepared data. Here, all metrics were hypothesised to deviate using Gaussian distribution $N(\mu, \sigma)$, and the range of resampled data was controlled between $\mu \pm k\sigma (k = 1, 1.5, 2)$. As a result, 300 sets of data were prepared in total, with 100 sets of data being generated for each participant group.

The prediction models were trained with the data generated using the Random Forest technique once again, and actual measured data was tested. Recall performance between k parameters was compared, and $k = 1.5$ produced the highest level of performance.

The test results are presented in Table 3.6 using the same format as the previous table, where all generated data for Tests 1–6 is employed as the set of trained data. The results show that for all participants in the AD group, AD was able to be predicted using both eyes. The data extensions may have compensated for the unbalanced data samples. However, incorrect predictions for MCI and NC groups increased, and some indices of performance decreased slightly. The number of predictions of the lower triangular matrix was sufficiently high, and thus the possibility of using prediction as a diagnostic procedure was confirmed.

Table 3.6 Classification results using Tests 1–6 with data extensions

Subject group	(1)	(2)	(3)	Recall	F1	Prediction pairs
(1) AD	8	0	0	1.00	0.64	ad-ad:4/4
(2) MCI	7	20	1	0.71	0.71	mci-*:13/14
(3) NC	2	5	7	0.50	0.64	nc-*:4/7
Precision	0.47	0.71	0.88	P-mean:0.69; R-mean:0.74		

Rate of lower triangular matrix: 0.98

Table 3.7 Classification results using Tests 1 and 2 with data extensions

Subject group	(1)	(2)	(3)	Recall	F1	Prediction pairs
(1)AD	8	0	0	1.00	0.64	ad-ad:4/4
(2)MCI	6	21	1	0.77	0.82	mci-*:12/14
(3)NC	3	2	9	0.64	0.75	nc-*:6/7
Precision	0.47	0.91	0.90	P-mean:0.76; R-mean:0.80		

Rate of lower triangular matrix: 0.98

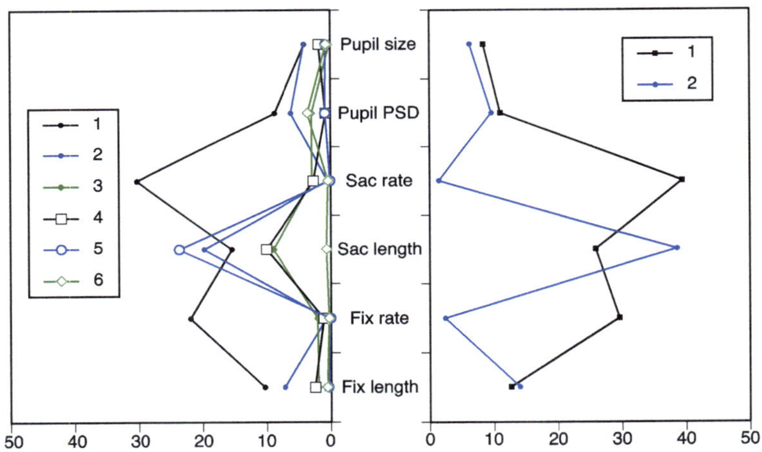

Fig. 3.6 Comparison of Gini coefficients of two conditions. Left side: Tests 1∼6. Right side: Tests 1∼2

In the previous section, features of metrics of patients in the early stages of consultation were suitable for prediction. A second set of results is summarised in Table 3.7, where features of Tests 1 and 2 are limited. In these results, NC group prediction performance improved, and some of the performance metrics showed the highest rates of correct prediction.

For a detailed analysis of performance, the contributions of features of observed metrics between the two models are compared in Tables 3.6 and 3.7. The degree of contribution of prediction as an index of importance (the Gini coefficient) is summarised in Fig. 3.6. The left side represents coefficients in Tests 1–6, and the right side represents coefficients in Tests 1–2. The figure suggests that the coefficients for Test 1, in particular saccade frequency, saccade length and fixation frequency, present a larger contribution than do the features of pupil responses. The coefficients on the left side suggest that most coefficients in Tests 3–6 were relatively small. These results confirm the importance of using metrics in the early stages of the consultation, as the changes between the initial cognitive load and the control session may represent the situation of individual participants.

3.4 Participant Classification Using Metrics of Oculo-Motors

3.4.1 Feature Extension

Additional features of oculo-motors, such as eye oscillations, are introduced to classification analysis in order to improve performance.

Two-dimensional frequency powers were extracted from eye movement oscillations, including fixations, in regard to the previous study [14]. Two factor metrics were calculated as a Promax solution in order to reduce the dimensions of the frequency powers. Though the two factor scores correlate with each other ($r = 0.77$), they are indicated for two directions of horizontal and vertical axes. The metric of eye oscillation of the three groups is compared, the mean for MCI is the highest power, and the mean for AD is the lowest, though there are no significant differences, however. The sampling rate of eye movement measurements was low, as deviations in frequency power may have been restricted.

3.4.2 Classification of Groups of Participants Using Features

The feature set of oculo-motors is extended to 11 dimensions, and the classification performance is examined using the Random Forest technique. Here, 11 features are summarised.

1. Pupillary change: mean pupil size [one dimension]
2. Frequency power of pupillary oscillation [15]: two frequency bands (0.5∼1.6 Hz C 1.6∼4.0 Hz) [12, 13] [two dimensions]
3. Eye movement features: saccade frequency, saccade length, fixation frequency, fixation duration [four dimensions]
4. Eye oscillation: two factor scores for each eye [four dimensions]

Once again, a data extension technique has been applied, as mentioned above, in order to compensate for the number of participants. Here, the training data set was generated using the Gaussian distribution $N(\mu, \sigma)$, and a range of resampled data was generated between $\mu \pm k\sigma (k = 1, 1.5, 2, 2.5, 3, 3.5)$. As a result, 300 sets of data were prepared in total, consisting of 100 sets for each participant group.

3.4.3 Classification Performance and Feature Selection

Classification performance using metrics of eye oscillations is examined. When all features of all tests are introduced (Tests 1∼6: 66 variables consisting of six tests × 11 dimensions), the F1 index decreases to 0.34, and all AD participants are not detected. The results of a comparison of performance are summarised in Table 3.8, with previous results used as a base line as indicated in the first line.

The Boruta technique [6] is introduced to select contributing variables from 66 variables, though a model using 30 variables nominated this way produces a lower level of performance (F1 = 0.32). The selected variable list shows most metrics of eye oscillations across the six tests, in addition to fixation and saccade information from Tests 1 and 2. Once again, these results suggest that the metrics of Tests 1 and 2 provide significant information.

In the results, the features in Tests 1 and 2 are focused on 22 variables in regard to the result of the previous model, as the F1 index increases slightly (F1=0.43). By reducing the number of features using the Boruta technique, 17 variables are selected, but the F1 index for this condition does not improve. Therefore, the Gini coefficient is used for feature selection instead of the Boruta technique. Variables with Gini coefficients are above 5 (Gini >5) are selected, and classification performance is evaluated using weight parameter k for data extension. The number of selected variables in Tests 10∼16 changes when the F1 indices for the three participant groups change. These changes are illustrated in Fig. 3.7 using open symbols. In two conditions, the AD group is not detected correctly, and the mean F1 index does not improve.

In addition, variables such as Gini coefficients over 10 (Gini >10) which contributed to performance to a greater extent are focused on, and the results are illustrated in Fig. 3.7 using solid symbols. The F1 index for the NC group improves, and the mean F1 value reaches 0.65. Classification performance of this condition is summarised in Table 3.9 as a contingency table. The number of participants is double since both eyes are being classified. Though the precision rate for the AD group and the recall rate for the NC group are lower, the mean F1 value as a measure of overall performance improves, so that in Table 3.8 this condition is ranked the highest. Also, the rate of the lower triangular matrix is sufficiently high, at 0.96.

Table 3.8 Comparison of classification performance

Introduced features in models	No. of variables	k	F1
Pupil and eye movements in Test 1∼2 (above model) [12]	12	1.5	0.60
All features in Tests 1∼6	66	1.5	0.34
Feature selection (Boruta) in Tests 1∼6	30	1.5	0.32
All features in Tests 1∼2	22	1.5	0.43
Feature selection (Boruta) in Tests 1∼2	17	1.5	0.42
Selected features (Gini >5) in Tests 1∼2	16	1.5	0.40
Selected features (Gini >10) in Tests 1∼2	5	1.5	0.65

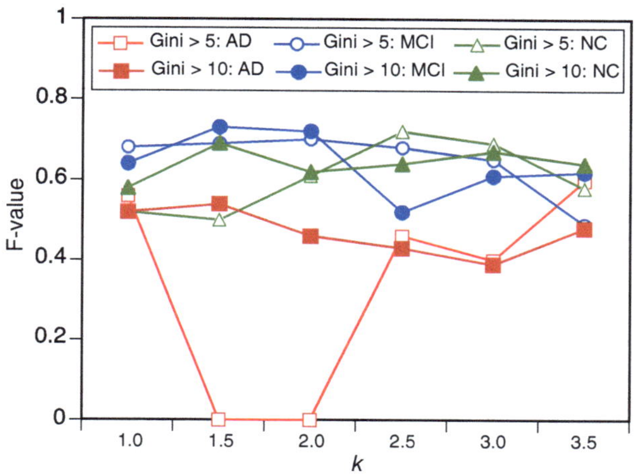

Fig. 3.7 Change of F1 index with feature selection using Gini coefficient and coefficient (k) in data extension

Table 3.9 Classification results using Tests 1∼2 (Gini >10)

Subject group	(1)	(2)	(3)	Recall	F1
(1) AD	7	1	0	0.88	0.54
(2) MCI	9	20	1	0.67	0.73
(3) NC	2	4	8	0.57	0.69
Precision	0.39	0.80	0.89	$(\bar{P})$:0.69; $(\bar{R})$:0.71	

Rate of lower triangular matrix: 0.96

Once again, participants' measurements taken during Tests 1 and 2, in the early stage of diagnosis, may provide stable responses.

During feature selection, the Boruta technique suggests that eye oscillations or eye movements contribute to the results of all tests. However, metrics of eye movements in Tests 1 and 2 are selected by the Gini coefficients of variables, such as saccade frequency, saccade length, fixation frequency and fixation duration. However, as pupil metrics are excluded, emotional factors may affect pupil response due to the level of cognitive ability or prevailing mental workload. Also, the effectiveness of data extension on the participant classification in a limited clinical survey is confirmed once again.

3.5 Summary

In this chapter, changes in pupil responses and eye movement of AD, MCI and NC participants during cognitive test such as the MMSE were measured, and the possibility of classification of participant groups was examined using their measured features. The following topics were discussed.

1. Several metrics of oculo-motors were extracted from observed eye images during six sessions of MMSE cognitive tests, and the differences in pupil size, pupillary oscillation, fixation, saccade and eye oscillation between participant groups were compared.
2. The number of participants in clinical surveys was limited, and a data extension technique was used to develop a model to classify groups of participants.
3. The Random Forest technique was employed for the classification procedure. Also, two feature selection procedures, the Boruta technique and the Gini coefficient, were introduced to improve performance. As a result, a model was generated using five variables which were Gini coefficients greater than 10, which provided better classification performance (the mean F1 was 0.65). The contribution of features of fixation and saccade information during controlled conditions was confirmed.

More effective feature selection and the improvement of classification performance will be subjects of our further study.

Acknowledgments This work was partially supported by the Polish Ministry of Science and Higher Education research grant NN518 405338 and by the International Collaboration Research Promotion Fund, School of Engineering, Tokyo Institute of Technology.

References

1. Aoki, K., Ngo, T.T., Mitsugami, I., Okura, F., Niwa, M., Makihara, Y., Yagi, Y., Kazui, H.: Early detection of lower MMSE scores in elderly based on dual-task gait. IEEE Access **7**, 40085–40094 (2019)
2. Beatty, J.: Task-evoked pupillary response, processing load, and the structure of processing resources. Psychol. Bull. **91**(2), 276–292 (1982)
3. Ebisawa, Y., Sugiura, M.: Influences of target and fixation point conditions on characteristics of visually guided voluntary saccade. J. Inst. Image Inform. Television Eng. **52**(11), 1730–1737 (1998)
4. Folstein, M.F., Folstein, S.E., McHugh, P.R.: MINI-MENTAL STATE—a practical method for grading the cognitive state of patients for the clinician. J. Psychiatr. Res. **12**, 189–198 (1975)
5. Kuhlmann, J., Böttcher, M. (eds.): Pupillography: Principles, Methods and Applications. W. Zuckschwerdt Verlag, Munchen, (1999)
6. Kursa, M.B., Rudnicki, W.R.: Feature selection with the boruta package. J. Stat. Softw. **36**(11), 1–13 (2010)
7. Nakayama, M., Katsukura, M.: Development of a system usability assessment procedure using oculo-motors for input operation. Univ. Access Inf. Soc. **10**(1), 51–68 (2011)
8. Nakayama, M., Nowak, W., Ishikawa, H., Asakawa, K., Ichibe, Y.: Discovering irregular pupil light responses to chromatic stimuli using waveform shapes of pupillograms. EURASIP J. Bioinform. Syst. Biology **#18**, 1–14 (2014)
9. Nasreddine, Z.S., Phillips, N.A., Bédirian, V., Charbonneau, S., Whitehead, V., Collin, I., Cummings, J.L., Chertkow, H.: The Montreal Cognitive Assessment, MoCA: A brief screening tool for mild cognitive impairment. J. Am. Geriatr. Soc. **53**, 695–699 (2005)
10. Nowak, W., Nakayama, M., Kręcicki, T., Hachoł, A.: Detection procedures for patients of Alzheimer's disease using waveform features of pupil light reflex in response to chromatic stimuli. EAI Endorsed Trans. Pervasive Health Technol. **6**, 1–11 (2020). E6

11. Nowak, W., Nakayama, M., Kręcicki, T., Trypka, E., Andrzejak, A., Hachoł, A.: Analysis for extracted features of pupil light reflex to chromatic stimuli in Alzheimer's patients. EAI Endorsed Trans. Pervasive Health Technol. **5**, 1–10 (2019). E4
12. Nowak, W., Nakayama, M., Trypka, E., Zarowska, A.: Classification of Alzheimer's disease patients using metric of oculo-motors. In: Proceedings of the Federated Conference on Computer Science and Information Systems (FedCSIS), pp. 403–407 (2021)
13. Peysakhovich, V., Causse, M., Scannella, S., Dehais, F.: Frequency analysis of a task-evoked pupillary response: luminance-independent measure of mental effort. Int. J. Psychophysiol. **97**, 30–37 (2015)
14. Schwetlick, L., Rothekegel, L.O.M., Trukenbrod, H.A., Engbert, R.: Modeling the effects of perisaccadic attention on gaze statistics during scene viewing. Commun. Biol. **3**(727), 1–11 (2020)
15. Ukai, K.: Pupil. In: Japan Society of Vision Science (ed.) SHIKAKU JYOUHOU SYORI HANDOBUKKU (Handbook of Visual Information Processing). Asakura Shoten, Tokyo, Japan (2001)

Chapter 4
Detection of Dementia Symptoms in Elderly Using Features of Binocular Pupil Light Reflexes to Chromatic Light Pulses on Either Eye

Abstract Procedures for calculating the probability of levels of dementia are developed using features of binocular pupil light reflex (PLR) to a chromatic light pulse on either eye. The PLR waveforms of 101 elderly participants, consisting of patients with Alzheimer's disease (AD) and mild cognitive impairment (MCI) or belonging to a normal control (NC) group, were measured under four experimental conditions. Three factor scores were calculated from the PLR waveform features for each response. Also, metrics of pupillary oscillation were introduced. The probability for the level of dementia of each participant was calculated using two types of regression functions for MCI+AD or AD patients. During function optimisation, some variables that represent asynchronous measurement and pupillary oscillation were selected, and the requirement of binocular measurement was confirmed. The results confirm the effectiveness of measuring features of ocular oscillation as a method of classifying patients. As the PLR responses of each eye were different during light pulse irradiations of the left or right eye, overall performance may be improved by using two independent classifications.

Chapter Co-authors: Minoru Nakayama, Wioletta Nowak, and Anna Zarowska

Originally published as follows:

Minoru Nakayama, Wioletta Nowak, Anna Zarowska, "Symptoms of Dementia in Elderly Persons Using Waveform Features of Pupil Light Reflex", LNBIP 471, pp. 91–107, March, 2023.

Minoru Nakayama, Wioletta Nowak, Anna Zarowska, "Prediction Procedure for Dementia Levels Based on Waveform Features of Binocular Pupil Light Reflex", Proc. ETRA 2023: 16:1–16:7, May, 2023.

Minoru Nakayama, Wioletta Nowak, Anna Zarowska, "Extracting Dementia Symptoms in Elderly Using Ocular Motor Activity during Pupil Light Reflexes to Chromatic Light Pulses on Either Eye", Proc. iV2024, pp. 75–80, October 2024.

M. Nakayama, W. Nowak, *Development of Diagnostic Procedures Using Features of Pupil Reactions*, Behaviormetrics: Quantitative Approaches to Human Behavior 21, https://doi.org/10.1007/978-981-95-4502-5_4

4.1 Introduction

Symptoms of cognitive function impairment are used to diagnose Alzheimer's disease (AD) and mild cognitive impairment (MCI). A major diagnostic procedure is called the Mini-Mental State Examination (MMSE) [13], which is based on a set of face-to-face clinical tests with verbal communication. These require participants to have sufficient communication skills, however. Therefore, a quicker and easier objective procedure should be developed using appropriate bio-makers. In order to aid the diagnostic process, oculo-motor indices are often analysed to detect symptoms of dementia [3, 30]. These procedures may provide useful additional diagnostic information, as some elderly people have auditory ability impairments. Features of eye movements are widely used to detect evidence of AD [34, 42, 47]. Some eye movements such as saccades are observed in response to the visual task. Features of eye movements have been used to introduce diagnostic procedures in the past [31, 53]. Participant workloads during PLR measurement should also be considered [40].

Another oculo-motor metric is pupil response, such as the pupil light reflex (PLR), which represents the dynamic characteristics of pupil behaviour [4, 28]. The study of conventional PLR activity suggests that as this activity represents the visual information processing of retinal stimuli and the ability to activate neural signal transfers [5, 15], it should be evaluated as an alternative means of diagnosing cognitive function impairment [3, 30]. Also, PLR responses based on melanopsin ganglion cells [17, 24, 56] can be applied to the study of Aged Macular Disease (AMD) and AD [7, 10, 17, 24], and the possibility of their use in diagnosing these diseases has been studied [36, 43, 44, 46]. A simple procedure to detect AMD and AD patients is required for medical and clinical staff who treat elderly people [45]. In a sense, a diagnostic procedure using ocular-motors may be an easy method, as it does not require verbal communication.

Some diagnostic procedures based on PLR analysis have been studied in order to distinguish possible patients from normal control (NC) groups using a single pupil response [43, 44, 46]. In considering the PLR mechanism, binocular responses to PLR waveforms should be evaluated in response to light pulses to either eye [58].

The authors have been conducting feasibility studies about how to conduct PLR observations at clinical institutions using a portable measuring system. During the current survey, additional elderly people were invited to participate, and their responses were analysed in order to examine the possibility of diagnostic prediction. PLR characteristics of both eyes in response to chromatic stimuli of either eye should be analysed carefully, in addition to the introduction of an index of pupillary oscillations which can serve as an additional feature [45]. Also, the possibility of detecting whether the level of dementia indicates AD or MCI should be examined.

In this chapter, the following points are addressed:

1. Features of PLRs for blue and red light pulses of the left and right eyes are compared, and the differences are extracted.

2. The ability of classifying participants as AD/MCI or normal control (NC) using MMSE scores and PLR features is evaluated.
3. The contribution of the participant's age is also examined.
4. A procedure to classify the level of dementia as AD, MCI and NC is developed, and the performance of this procedure is examined.
5. A prediction procedure for detecting dementia of patients using features of PLR waveforms of both eyes evoked by light pulses on either eye is examined.
6. The features of frequency powers of eye centre positions and pupillary oscillations of groups of AD, MCI and NC participants are compared.
7. Prediction performance of both sets of irradiated eyes using features of PLR waveform shapes and oscillation frequency powers of eye and pupillary changes is evaluated.

4.2 Related Work

4.2.1 Diagnostic for Alzheimer's Disease (AD)

The study of Alzheimer's disease in clinical and pathological approaches has a long history [6, 12]. Early diagnosis and appropriate treatment of the disease are required [49, 51]. The diagnostic and testing procedures currently in use have been studied, and their accuracy has often been discussed [2, 9]. Though various clinical aspects of AD patients or patients with dementia, such as mild cognitive impairment (MCI), have been studied, the typical features that present AD symptoms have not yet been examined. Major diagnostic tools are almost always based on clinical and pathological measurement such as brain imaging and certain biomarkers [26, 32], while clinical surveillance is frequently preferred using medical consultations.

Even so, clinical doctors have to diagnose using observations such as the MMSE (Mini-Mental State Examination) question inventory [13] or other procedures. As MMSE is used by observers to mark conditions during subject observation, the validity of the results depends on the activity of the targeted participant. Some improved consultation procedures are also being developed to enhance accuracy [54].

However, the diagnostic procedure is based on clinical observations using verbal communication. If the participants have sensor impairments, it may not be possible to conduct these types of tests. In another procedural approach to diagnosis, participant's behavioural responses, acoustic features of their spoken voice [20, 27, 57], gait features [1] and oculo-motors [43, 44] have been analysed technically. Some new measurements using vision functions are also being studied to introduce new diagnostic approaches [10, 30, 55].

In particular, eye pupil reaction such as the pupil light reflex (PLR) is often referred to as one of the biomarkers for detecting AD patients since melanopsin-related ganglion cells (ipRGCs) have been discovered [8, 16, 18, 24]. The detailed mechanism of the PLR and diagnostic procedures for detecting cognitive impairment based on ipRGCs and ipRGC-based PLRs have been studied [10, 56]. Regarding the possibility of diagnosing AD patients, most conventional observations were conducted using a flash light and employing white light [14] of longer wavelengths, such as 585 nm [5], 820 nm [14] or other wavelengths [10, 16]. The observation procedures and the characteristics extracted from PLRs have been discussed regularly. Some characteristics of PLR waveforms are extracted and compared with light wavelengths of colours such as blue or red [11, 35, 58] as suggested in previous works [22–24]. Though some PLR-based diagnostic prediction procedures using features of the waveforms are proposed [43, 44], key features of PLRs and the appropriate procedures have not yet been extracted. As the previous studies have produced various successful assessments, feature extraction approaches to ipRGC-based PLRs will be discussed in this work, with the most appropriate observation procedure for a PLR-based diagnostic prediction procedures also being considered.

4.2.2 *PLR Mechanism*

Most of these procedures are based on the performance of signal transfer from retinal ganglion cells to the pretectal area via the optic chiasm, while the Edinger-Westphal (E-W) nucleus also plays a major role in PLR [25, 33, 58]. In regard to the transfer system, assessment is focused mainly on retinal activated signals from ipRGCs, which are activated by chromatic stimuli and are transmitted to the pretectum while evoking a pupil constriction, as shown in Fig. 4.1 [33, 38]. Some impairment of the transfer path and the E-W Nucleus, such as neuronal loss, may influence AD patients [10, 52]. Overall performance of the synaptic signal transfer may not be evaluated individually.

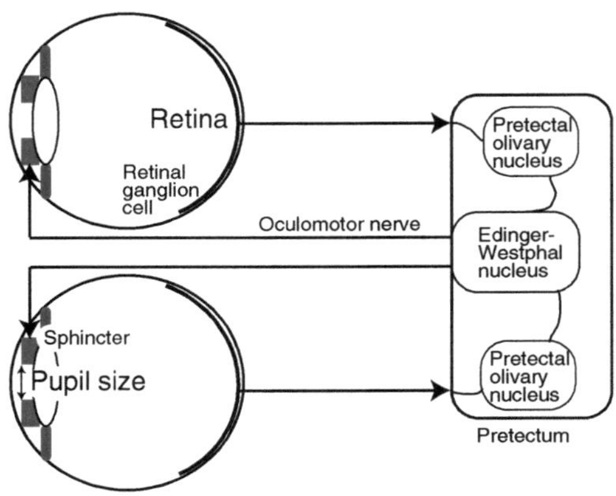

Fig. 4.1 Diagram of the pupil light reflex

In the case of a proposed diagnostic procedure for Aged-Related Macular Disease (AMD) which also influences synaptic signal transfer, asynchronous PLR appearances between the eyes were focused on and evaluated quantitatively using waveform features [36, 50]. It may be possible to consider using this type of experimental protocol to assess procedures for detecting symptoms in AD patients. These two diseases influence the extent of retinal damage or the contribution of photoreceptors, and because they serve as pathways for signal transfer [25], they can be used as a means of simplifying AMD diagnosis [36]. In regard to the diagnostic issues, observation of these conditions should be focused on using the responses of each eye, even when the reactions of both eyes are measured synchronously. In most studies, the difference between the eyes is not considered since both pupils and eyes are thought of as changing synchronously, such as during consensual light response [58]. Also, similar processing can even be employed during diagnostic procedures that use PLR responses. This aspect should be considered using binocular synchronous observation in the analysis. For example, the following experiment may be introduced so that the PLR responses of both eyes are evaluated separately when either eye is irradiated by short- or long-wavelength light pulses. Therefore, the performance of the retinal signal transfer pathway of either eye and the reflex response signal transfer pathways of both eyes could be evaluated when a light pulse is presented to either eye. The following experiment introduces binocular pupillary observations in response to light pulses on either eye as a means of inducing chromatic stimulus.

4.2.3 Participant Factor

Most AD patients and persons presenting symptoms of dementia, including MCI, are elderly, and almost all are receiving medication to treat health problems. In addition to individual differences including cognitive performance [19], the pupil size generally decreases with age due to senile miosis. Cataract of the lens may also influence sensitivity to PLRs. In order to create a detection procedure for AD patients, a comparison of patients and normal control (NC) subjects is required initially. Even NC persons may exert some influence over their PLR responses.

Fundamentally, recruiting AD and MCI patient volunteers may not be easy once the experimental procedure is ethically approved [44]. Even a clinical doctor may only ask a limited number of candidates to participate. Therefore, most experiments have to be conducted by employing a small number of patients. If the number of participants grows, the condition of each individual with dementia and their health issues should be considered carefully during the evaluation of the results of experiments. Also, measured pupillary data should be carefully processed since participants may not be able to control blink and eyelid lift during observation. At minimum, the age and cognitive ability test scores of participants are required, but the deviations caused by individual differences need to be considered.

In order to invite elderly people to participate in an observation experiment which contained a balance between patients and NC group participants, family members of patients were encouraged to join so that a sufficient number of observations of additional elderly people could be made.

4.3 Method

The experiment was designed to evaluate the selective responses of irradiated ganglion cells on the retina and the PLR mechanism [39]. A simple PLR diagram is shown in Fig. 4.2. Signal pulses of light activated ganglion cells on either eye are transferred to the pretectum via the optic nerve, and the reflex response controls the sphincter of both pupils via the oculo-motor nerve [33].

4.3.1 Experimental Procedure

Participants were introduced to a temporary dark space, where the following five experimental sessions were conducted for 10 seconds each.

1. Condition 1: Control session without light pulses
2. Condition 2: Blue light pulse to the right eye
3. Condition 3: Blue light pulse to the left eye
4. Condition 4: Red light pulse to the right eye
5. Condition 5: Red light pulse to the left eye

The experiment is designed to study the influence of light pulses on synaptic connections between both eyes in response to light pulses to either eye. Light pulses

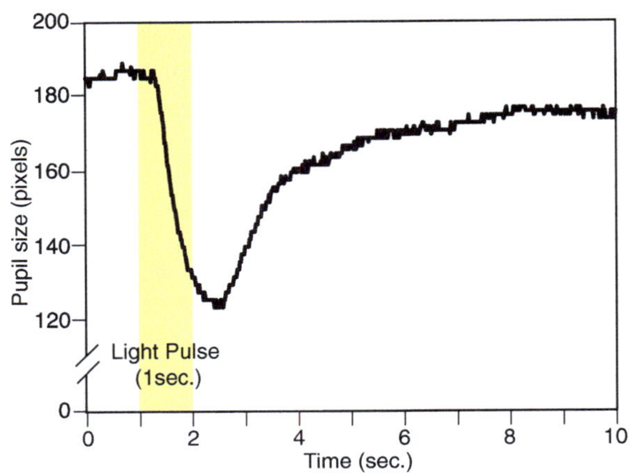

Fig. 4.2 An example of a PLR waveform (yellow area indicates a 1 second light pulse)

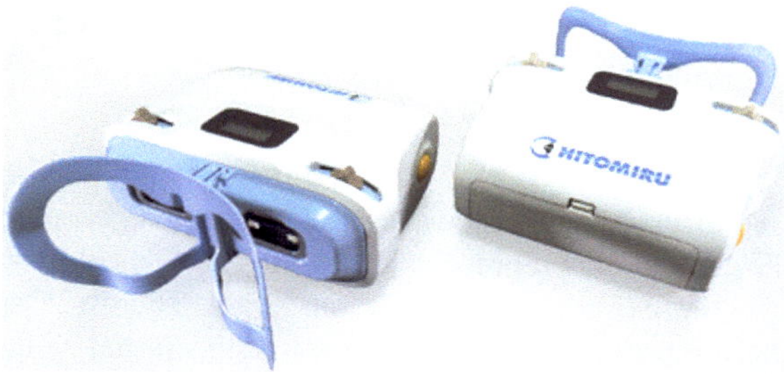

Fig. 4.3 Equipment to observe pupillary changes

transfer from retinal ganglion cells on the irradiated eye to sphincters of both eyes via the Edinger-Westphal nucleus, as mentioned above. The processes of miosis and restoration were observed in all four sessions. A short relaxation break was inserted between each session.

The size of pupil responses was measured in pixels at 60 Hz using equipment with blue and red light sources, as shown in Fig. 4.3 (URATANI, HITOMIRU). The light sources were blue (469 nm, 14.3 cd/m^2, 6.5lx) and red (625 nm, 12.3 cd/m^2, 10.5lx). The size of both pupils during all conditions was measured. Blink artefacts were removed manually after measurement.

The experiment was conducted by a clinical physician at a medical institution, and the procedure was approved by an ethics committee of Osaka Kawasaki Rehabilitation University.

4.3.2 Participants

Valid data was obtained from 101 participants, of which 66 were female and 35 were male. Their mean age was 78.5, and the SD (standard deviation) was 8.9 years. Participants at a medical institute were selected, and an MMSE test was conducted. The results were classified into three groups according to MMSE scores. These were AD (Alzheimer's disease, with MMSE <=23), MCI (mild cognitive impairment, with MMSE <=27) and others, whose conditions were NC (Normal Control). The distribution was as follows:

- AD: 31 (F:21, M:10), mean age: 83.0, SD: 6.3 years
- MCI: 9 (F:5, M:4), mean age: 82.1, SD: 6.3 years
- NC: 61 (F:40, M:21), mean age: 75.6, SD: 9.2 years

As the age of participants may influence their condition, four age levels were created: less than 66 years old (0), 66–75 years old (1), 76–85 years old (2), and above 85 years old (3). Though participants were elderly and might have some health problems, these points were not considered during the following analysis.

4.4 Regression Prediction Using Waveform Features of PLRs

4.4.1 PLR Waveforms

An example of PLR waveforms for an NC participant is shown in Fig. 4.4. The horizontal axis represents time, and the vertical axis represents pupil size in pixels for the experimental conditions 2∼5. There are some differences in pupil size between the left and right eyes at the initial point. The legend “BR-R” means Right pupil response when Blue light irradiates the Right eye, and also “RL-R” means Right pupil response when Red light irradiates the Left eye. Also, levels of contraction are different between conditions, as in the previous work that presented PLR responses to blue or red light pulses [24]. Another example of an AD patient is shown in Fig. 4.5. Some typical features are observed, such as deviation during the restoration process after constriction. Several features of waveforms were extracted in order to compare groups of participants in relation to the previous study [43], as shown in Table 4.1.

The first hypothesis is that there is a feature difference between the left and right eyes when light pulses are directed at either eye. The hypothesis was examined using a t-test of features of both eyes, such as between the irradiated eye and the non-irradiated eye. The features were extracted from standardised waveforms in order to reduce the potential differences.

In the results of the test, there are no significant differences in any of the features. There were a few exceptions, but the results did not coincide with the results of either colour of light pulse.

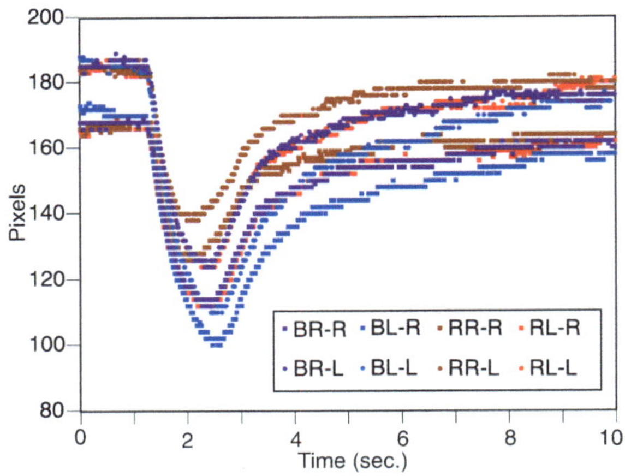

Fig. 4.4 Examples of PLR of both eyes for four conditions (NC participant, 76 yo, M), categories: [light colour][irradiated eye]-[observed eye]

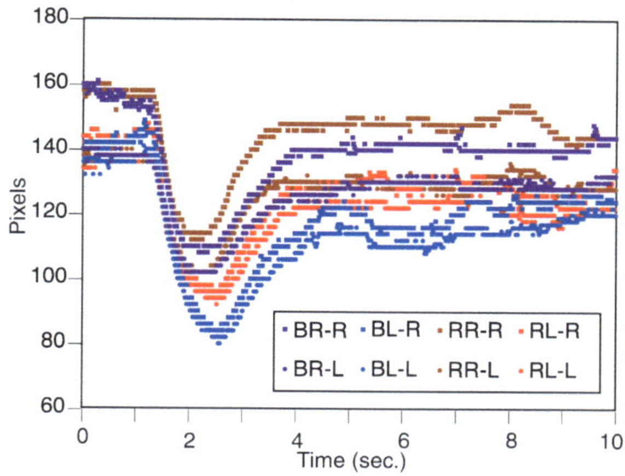

Fig. 4.5 Examples of PLR of both eyes for four conditions (AD participant, 87 yo, F), categories: [light colour][irradiated eye]-[observed eye]

Table 4.1 Features of PLR

Variables	Definitions
RA	Relative amplitude of miosis
t_min	Time at minimum size
diff_min	Minimum differential of size
t_diff_min	Time at minimum differential
diff_max	Maximum differential of size
t_diff_max	Time at maximum differential
diff2_min	Minimum acceleration
t_diff2_min	Time at minimum acceleration
diff2_max	Maximum acceleration
t_diff2_max	Time at maximum acceleration

4.4.2 Regression Analysis Based on Factor Scores

Since every feature includes measurement errors and individual differences, the latent factors are extracted using factor analysis according to the method used in the previous study [43].

The results of factor analysis are shown as a factor loading matrix in Table 4.2. In this work, three factors are employed, and the overall contribution ratio of the factors is 45.5%. Factor 1 represents the differential rate and acceleration of pupillary change, Factor 2 represents the features of contraction such as relative amplitude and its time, and Factor 3 represents the times for the differential rates and acceleration, as mentioned in Factor 1.

Three factor scores are calculated using the factor loading matrix. When the scores of both eyes are compared, there are also no significant differences.

The factor scores for experimental conditions are summarised and compared in Fig. 4.6. Changes in Factor-2 scores suggest a continuous decrease according to the experimental conditions. Also, there are significant differences in the three factor

Table 4.2 Factor loading matrix for PLR features

Variables	Factor 1	Factor 2	Factor 3
diff_min	**0.87**	−0.13	0.09
diff2_min	**0.76**	0.06	0.16
diff2_max	**−0.83**	−0.17	0.22
diff_max	−0.36	0.08	0.15
RA	−0.24	**0.78**	−0.09
t_min	0.22	**0.73**	0.14
t_diff2_min	−0.13	−0.00	**0.49**
t_diff_min	−0.05	−0.03	0.36
t_diff_max	−0.11	0.23	0.36
t_diff2_max	0.06	0.07	0.30

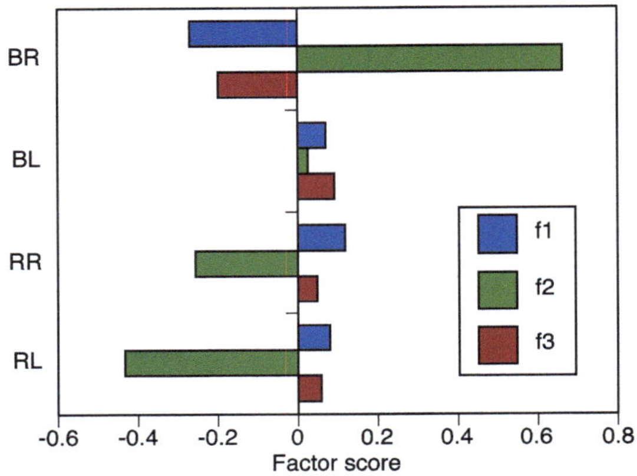

Fig. 4.6 Comparison of factor scores between stimuli (f1∼f3: factor scores for factors 1∼3)

scores for blue and red stimuli. Within each colour stimulus condition, there are significant differences in the three factor scores for blue light pulses and significant differences in Factor-2 scores for red light pulses ($t(402) = 2.13$, $p < 0.05$). The differences between sessions using the same colour condition should be considered, and in particular the differences for blue light should be evaluated separately.

In addition, the factor of age level on factor scores is examined using two-way ANOVA of participant groups and age levels. Though the factor for the participant groups is not significant, the factor for age level is significant for Factor-1 scores ($F(3, 801) = 19.9$, $p < 0.01$), and the interaction between the two factors (age and participant group) is also significant ($F(2, 801) = 9.4$, $p < 0.01$). Therefore, the factor of age may affect the differential and the acceleration of pupillary change, as presented in Factor-1 scores.

The influence of a subject's level of dementia on PLR features was not confirmed in the above analysis. The factors of stimulus light wavelength and the age of the patient were significant and are the major components of the deviation. In order to examine the effectiveness of the extracted features for a prediction of cognitive function impairment, an estimation procedure using a logistic analysis which had been introduced in a previous study [43] was conducted. Here, both MCI

Table 4.3 Prediction models using factor scores of PLRs

Model	Variables	AUC
1	Nine factor scores	0.77
2	Nine factor scores + age group	0.84
3	Nine factor scores + age	0.84
4	Selected variables: five factor scores + age group	0.84

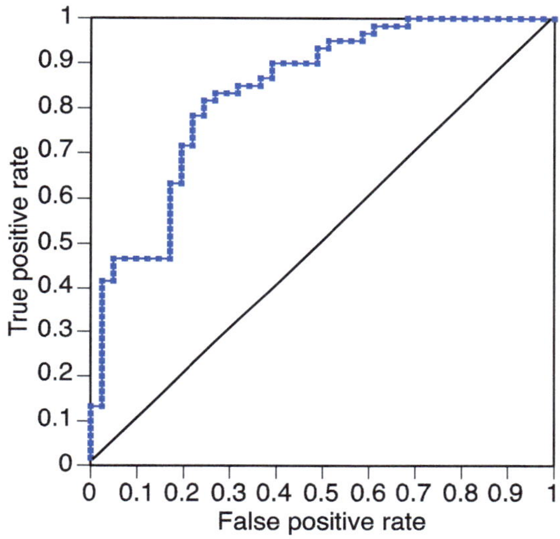

Fig. 4.7 ROC for Model 2 (AUC=0.84)

and AD patients are merged as the "AD+MCI" group since the number of MCI participants is limited. The probability of cognitive impairment is calculated using factor scores for each participant. As there are no significant differences between eyes, averaged features of responses of both eyes are employed. In considering the differences between session stimuli, two sets of features for blue light conditions and averaged features for red light conditions are introduced, for a total of nine variables altogether.

Table 4.3 shows a summary of several prediction models and AUC (area under the curve) as an index of accuracy of binary classification for a ROC (Receiver Operating Characteristic) curve. Since the threshold for classification may depend on the diagnostic policy, such as reducing the false positive rate, accuracy is evaluated using AUC. An example of ROC for Model 2 is illustrated in Fig. 4.7. Model 1 consists of nine factor scores, and Models 2 and 3 include age level or age. Model 4 employs significant contributing variables using a step-wise selection technique. Participant's age information aids classification performance.

Probabilities for the classification of cognitive impairment based on Model 2 are calculated, and the relationships between MMSE scores and the probabilities are summarised in Fig. 4.8. The horizontal axis represents MMSE scores, and the vertical axis represents probability. Participants who were tested using MMSE

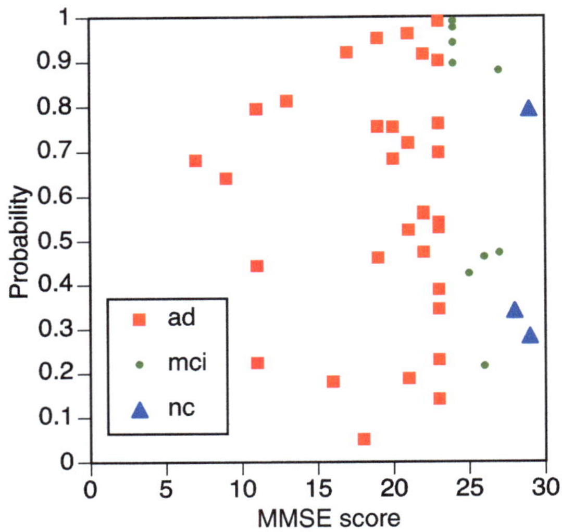

Fig. 4.8 Relationship between MMSE scores and computed probabilities

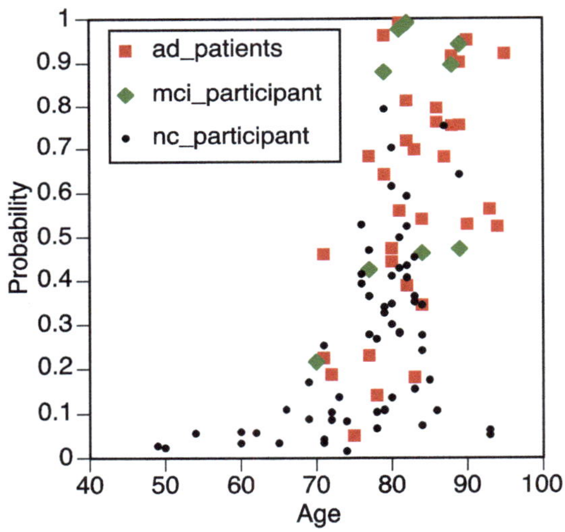

Fig. 4.9 Change in probability according to participant's age

are plotted in the figure according to their participant group, AD, MCI or NC. Confirmation of the contribution of a participant's age is shown in Fig. 4.9, where the horizontal axis represents age. In this figure, cognitive impairment can be observed in subjects over 70 years old, and the probability increases markedly from around 70 onward. When the threshold for AD+MCI is set to 0.5, 80% of participants are classified correctly. The AUC is 0.84 as shown in Table 4.3.

Mean probabilities for the groups of AD+MCI and NC by age level are summarised in Fig. 4.10, in order to evaluate the contribution of age level. Overall, mean probabilities increase with the age level. In particular, the probability of

Fig. 4.10 Comparison of mean probabilities between AD+MCI and NC groups

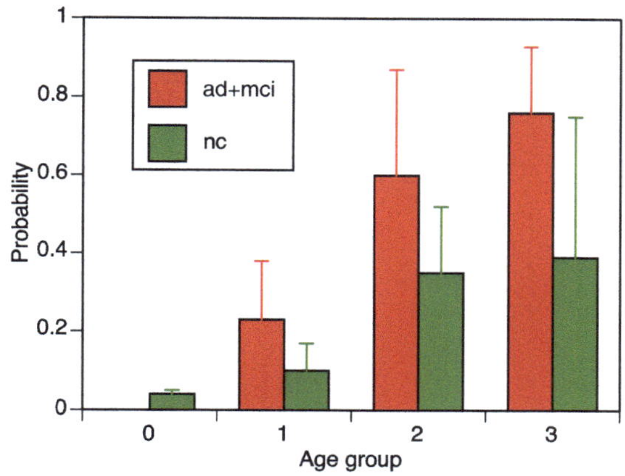

AD+MCI increases over age 75, while mean probability of NC remains under 0.5. Model 4 in Table 4.3 was generated using a variable selection procedure. All five selected variables are factor scores for blue light pulses during two test sessions, and factor scores for red light pulses were not used. The AUC fitting index is comparable with the values of the other functions. This suggests the possibility that prediction can be made using responses to blue light pulses. A more detailed examination of this topic will be summarised in the following Discussion section.

4.4.3 Discussion

In the first hypothesis, cognitive impairment may be affected by the influence of the oculo-motor nerve, which connects retinal ganglion cells to the pretectal area on the synaptic path. However, no significant differences in the extracted features were observed during several chromatic light pulses, though there were some differences between the experimental sessions, as shown in Fig. 4.4. One of the possible reasons is the dependence on the accuracy of measurement, because feature extraction is based on point estimation. In particular, the small pupils of senior citizens may influence the measurement of temporal change of pupil size. Another problem may be that the extracted features focused on the constriction phase of PLRs without allowing for the restoration phase which follows. The difference in PLR between the left and right eyes should be measured carefully in considering the above points. During the main analysis, the AD+MCI group was set as the specific target for prediction. As a diagnostic application, the level of cognitive impairment needs to be able to be estimated using features of PLRs if it is to be effective. If it is possible, AD and MCI should be classified using weighted levels.

The factor scores for blue but not red light pulses were selected once more for use in a regression model, following a step-wise procedure. The dominance of blue light pulses for prediction was confirmed in a previous study [43, 44]. The possible

Table 4.4 Correlation coefficients for PLR waveforms between left and right eyes

Condition	1.0–2.5s	2.5–3.5s	3.5–10s	0.0–10s
BR:NC	0.99	0.98	**0.84**	0.97
BR:MCI+AD	0.99	0.97	**0.87**	0.97
BL:NC	1.00	0.99	**0.90**	0.98
BL:MCI+AD	1.00	0.99	**0.85**	0.97
RR:NC	0.99	0.97	**0.82**	0.96
RR:MCI+AD	0.99	0.96	**0.73**	0.97
RL:NC	0.99	0.96	**0.83**	0.96
RL:MCI+AD	0.99	0.96	**0.78**	0.96

reason for this may be based on the first hypothesis, which could not be confirmed according to the above evaluation, however. Therefore, a more detailed analysis needs to be conducted.

During the above waveform analyses, most features of both eyes were extracted from the constriction phase between 1.0 and 2.5 s. As the initial hypothesis states, the differences between eyes, such as a direct irradiated eye and a non-irradiated eye, were examined using a statistical mean differential test. In the results, there was no significant difference, and the hypothesis was rejected. However, some differences in waveform shapes are observed in Fig. 4.5, which illustrates the responses of an AD patient. Here, a more detailed analysis is tried using a simple mutual correlation analysis of two PLR waveforms for the left and right eyes. If the two eyes react synchronously, the correlation coefficient approaches 1. In considering the reaction phases such as constriction and the two stages of restoration, correlation coefficients are calculated and summarised across three phases for each stimulus condition in Table 4.4.

The overall correlation coefficients are presented in the right-hand column, and nearly all approach the value of 1 (0.96∼0.98). During the initial phases (1.0∼2.5 s and 2.5∼3.5 s), the coefficients approach 1 since the dynamical reaction occurred simultaneously during constriction and early restoration phases. The coefficients decrease to less than 0.9 during late restoration phase (3.5∼10 s), however. These coefficients suggest asynchronous responses between the two eyes. In particular, coefficients for MCI-AD participants are less than 0.8 when red light pulses are used. This suggests that differences in response of some patients may have occurred as part of the conventional light response mechanism. Though pupil responses during the second stage of restoration are studied as a post-illumination pupil response (PIPR) [21, 58], the synchronicity between the two eyes is not discussed sufficiently. As the phase difference appeared after the constriction phase, the reaction mechanism should be discussed carefully once more if the internal information processing for single-eye irradiation with short- and long-wavelength light-influenced responses. In regard to the hypothesis in this work, the details of this phenomenon could be resolved by referring to previous studies. Also, waveform features for the late restoration phase should be considered in the diagnostic procedure.

In addition to the above issues, there are some obvious limitations to this study. The first point is the accuracy of classification of patients. During the examination, the diagnostic category is only based on MMSE scores measured during consultations with a medical doctor. The degree of disease progress should be evaluated objectively. Also, the NC group may include some people who are patients. In this study, other metrics of cognitive functional ability have been observed, such as VSRAD (Voxel-Based Specific Regional analysis system for Alzheimer's disease), HDS-R (Hasegawa's Dementia Scale-Revised) and MoCA-J (a Japanese version of the Montreal Cognitive Assessment test). The development of an alternative diagnostic procedure that considers the level of cognitive function together with these metrics will be a subject of our further study. Even the design of the experimental conditions may have been insufficient. PLR responses were only observed for 10 seconds under certain conditions following a low intensity 1-second light pulse. Usually, observations should be conducted using at least two levels of light intensity. The second restoration phase after light pulse needs to be recorded for over 30 seconds in order to properly examine pupillary dynamics.

A revised experimental design that considers these limitations should be discussed. This will also be a subject of our further studies.

4.5 Regression Model Improvement Using Additional Features

In another experiment, metrics of pupillary oscillations during the rest condition were studied in order to evaluate their contribution to a diagnostic procedure for dementia [45]. This information can be extracted from Condition 1 in the experimental procedure described above. The possibility of detecting whether the level of dementia indicates AD or MCI should be examined using a set of features which includes pupil oscillations [41].

4.5.1 Frequency Powers of Pupillary Oscillation

The frequency power spectra of pupillary oscillations during the rest condition for the three groups are summarised in Fig. 4.11. The horizontal axis represents frequency up to 2.0 Hz, and the vertical axis represents power spectrum density (PSD). The responses of both eyes are summarised as independent eyes. PSDs are summarised as the three mean powers, such as ff1: less than 1.6 Hz, ff2: 1.6$\sim$2.0 Hz and ff3: 0.35$\sim$0.7 Hz, and were recorded using the same method as in the previous study [48]. In comparing the mean PSDs of the three levels of participants using one-way ANOVA and sub-level t-tests, the means of ff1 and ff3 for NC are significantly higher than the means for the MCI and AD groups ($p < 0.05$). These mean PSDs (ff1$\sim$ff3) are added as pupillary features of individual participants.

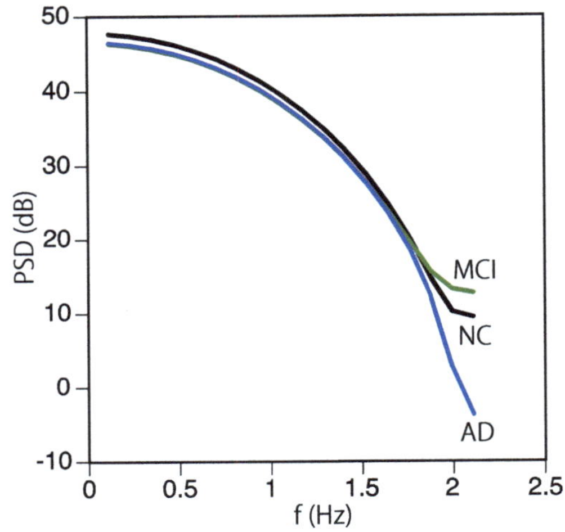

Fig. 4.11 Mean Frequency Power (PSD) for the three groups

4.5.2 *Regression Prediction for Cognitive Ability*

Participant's levels of cognitive ability are predicted using a logistic regression function with binocular features of PLRs, as follows. Initial probability sets the value for patients (MCI or AD) as 1 and 0 for non-patient (NC). Therefore, the predicted value indicates the probability for each participant.

- Model 1: Sum (R+L) and differential (R-L) of binocular features
  Enhancement of features using a summation of both feature values and differentials between the two eyes are employed for the prediction. Three factor scores in four conditions, (2) $\sim$ (5) [two sets $\times$ 3 $\times$ 4: 24 variables], and three PSDs [two sets $\times$ 3: 6 variables] and participant's age were used, and the total number of variables is 31.
- Model 2: All features of both eyes as independent features
  All features of the left and right eyes are introduced, and the total number of variables is also 31.

Two types of binary classification, such as NC vs. MCI+AD or NC+MCI vs. AD, are conducted using the above two sets of features, and the probabilities of each participant being in either class are calculated. The results of regression analysis are summarised in Table 4.5. Models for extracting NC participants are noted as "1n" and "2n", and models for predicting AD participants are noted as "1a" and "2a". Model 0 shows performance in a previous study as a baseline and employs mean features of both eyes.

In order to optimise the function, a variable selection procedure such as a stepwise method is applied, and contributing variables, which are called "1ns" and "2ns" and also "1as" and "2as", are extracted. The selected variables are noted using symbols such as the first "b" which represents a blue light pulse, the second symbol

Table 4.5 Logistic regression models for MCI and AD with NC groups

Model name	Variables of extracted features	No. of variables	AIC	ACC1	ACC2
0	Three factor scores in three conditions + age level [37]	9	118.4	0.84	0.68
1n	(R+L)(R-L) [3 factor scores x 4 conditions + 3 PSDs of Pupil] + age	31	120.7	0.89	0.67
1ns	Age, brd_f3, bls_f3, bls_f1	4	106.9	0.80	0.76
1a	The same condition for 1n [AD vs. MCI+NC]	31	118.9	0.95	0.66
1as	Age, brd_f3, bls_f3, s_ff2	4	104.3	0.79	0.77
2n	(R)(L)[3 factor scores x 4 conditions + 3 PSDs of Pupil] + age	31	120.7	0.90	0.66
2ns	Age, brr_f1, brl_f3, bll_f1, bll_f3	5	98.4	0.78	0.77
2a	The same condition for 2n [AD vs. MCI+NC]	31	118.9	0.93	0.64
2as	Age, bll_f3, brl_f3	3	108.0	0.73	0.72

ACC1: prediction accuracy based on entire data set including target at $th = 0.5$
ACC2: prediction accuracy of subject leave-one-out cross-validation (LOOCV) without a target at $th = 0.5$
Model notations: "n" is the classification of the "NC" group and "a" is the classification of the "AD" group, and "s" is the application of variable selections.
Variable notations: "brd_f3" means Factor-3 score for **bl**ue light irradiated **r**ight eyes, and the **d**ifference or **s**ummation of both eyes.
"brr" means blue light irradiation of the right eye, and the response of the **r**ight eye. "ff2" represents the pupil oscillation frequency power at 1.6~2.0 Hz.

which represents the irradiated eye (left or right) and a third symbol which shows variables as a sum or differential in Model 1 and represents the variables of observed eye (left or right) in Model 2. Additional variables, "ff1" ~ "ff3" are mean PSDs. The variable "brd_f3" represents the differentials of the two eyes using F3 scores. Another variable "brl_f3" represents left eye F3 scores, while blue light pulses irradiate the right eye. The variables of a cross-responded factor scores contribute to the models significantly.

The model fitness is compared using AIC. Two types of prediction accuracy were used. ACC1 is based on the entire data set, and ACC2 uses subject leave-one-out cross-validation (LOOCV). These accuracies are calculated using a threshold of 0.5 for the two-level classification, with variable selection in Tables 4.6 and 4.7 as examples. Also, classification performance using ROC curves is summarised in Fig. 4.12, and the AUCs are 0.88 for the model "2ns" and 0.78 for the model "2as".

Most AIC values for the two models are comparable or less than the values in the previous model (118.4). Though the number of variables increased for models without variable selection, the AIC values are comparable since the fitness of the model may have improved. Also, the AICs are reduced by the model when variable selection is applied. ACC1 may show the fitness of the observed data, and ACC2

Table 4.6 Prediction for NC using variable selection (Model "2ns")

$th = 0.5$	NC	MCI+AD	Sum
NC	50	13	63
MCI+AD	10	28	38
N of group	60	41	101

Table 4.7 Prediction for AD using variable selection (Model "1as")

$th = 0.5$	NC+MCI	AD	Sum
NC+MCI	63	17	80
AD	6	15	21
N of group	69	32	101

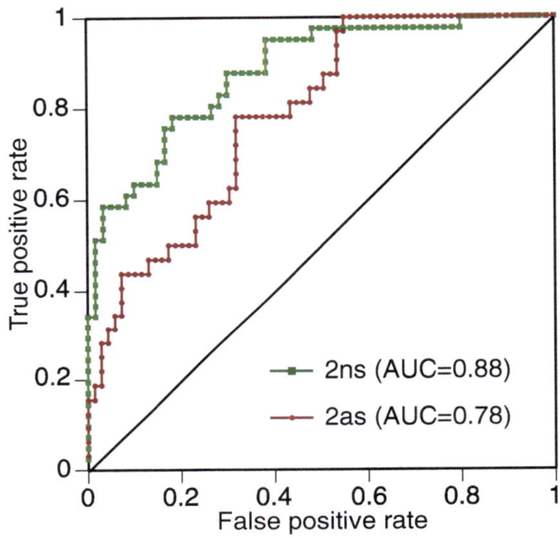

Fig. 4.12 ROC curves: Model "2ns" and Model "2as"

shows prediction performance. When the variable selection technique is applied to the models, the values of ACC2 are improved while the values of ACC1 decrease.

All selected variables are features when blue light pulses irradiate either eye, and no features based on red light pulse are selected. The selected factor scores are based on constriction amplitude (Factor 1) and constriction timing and acceleration (Factor 3). In particular, models that detect AD participants ("1as" and "2as") employ variables for Factor 3. Pupillary oscillation "ff2" is also selected for Model "1as". The factor of age contributes to all models consistently. In order to maximise ACC1, more variables were selected, but fewer variables were selected to maximise ACC2.

These results suggest the possibility that PLR features for blue light pulses can be used to predict cognitive performance, though the performance needs to be improved. The effectiveness of differentials for the responses of each eye was not confirmed sufficiently when compared with differentials using the features of both eyes.

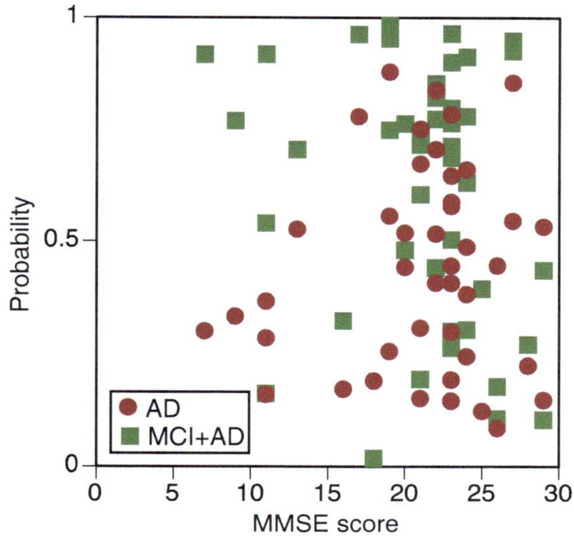

Fig. 4.13 Probabilities for MCI and AD using MMSE scores ("2ns" and "2as")

4.5.3 Possibility of Individual Prediction

As mentioned earlier, some of the participants were examined by a medical doctor in order to measure MMSE scores. Out of the 44 participants examined 41 were diagnosed as having MCI or AD. The probability of prediction using the models "2ns" for MCI+AD and "2as" for AD is summarised in Fig. 4.13.

Most MMSE scores are distributed around 20∼27, including the borderline cases of AD and MCI. The highest predicted probabilities are also distributed around the area of the graph with the most patients where MMSE scores are less than 28 and the deviation of the scores is large. When a threshold for detecting AD or MCI patients is set to 0.5, a certain number of patients may be detected using the model for MCI+AD, but only some detection of AD patients occurs using the model for AD only. Low MMSE scores may be indicative of patients with an acute level of illness, but the number of samples is small and sparse, so that the probability of calculation may not be measurable.

Practically, a diagnostic procedure is required to classify patients using the three levels NC, MCI and AD. In this work, two binary functions used to calculate probability were applied to this issue. Figure 4.14 shows a scatter gram with two probabilities, where the horizontal axis represents the probability of MCI+AD and the vertical axis represents the probability of AD. This figure includes NC participants. The two-dimensional region is divided into four sub-regions by two thresholds of probability of 0.5 each. Many patients are classified into high probability sub-regions, but some exceptions are classified into one low probability sub-region. Also, most NC participants are classified into the low probability sub-region, and a few are assigned to the high probability sub-regions. Some of the

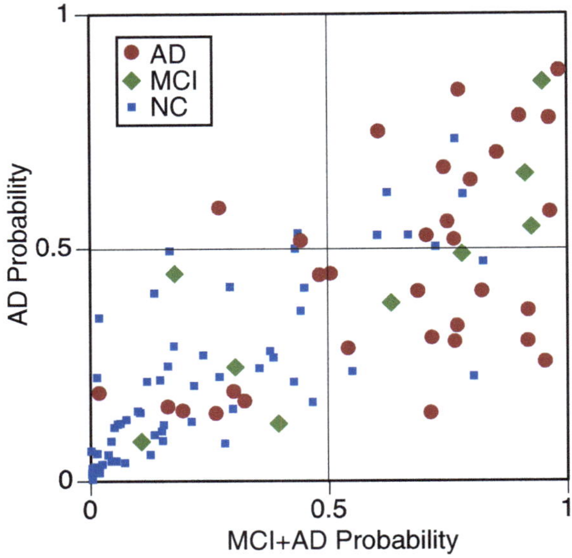

Fig. 4.14 Scatter gram of probabilities for MCI and AD ("2ns" and "2as")

NC participants are classified as patients. Though most NC group participants have not been examined by a doctor or taken an MMSE test, these participants may be included in the prediction of diseased patients using pupil light responses only. Therefore, patients who may possibly have AD or MCI disease can be diagnosed, as their plots fall into the high probability sub-regions. In order to determine further whether this technique would be a suitable diagnostic procedure, the accuracy should be improved.

4.5.4 Discussion

The frequency power of pupillary oscillations of participants during MMSE examinations contributed to detection performance of AD patients in the previous study [45]. In regard to this study, PSDs in the resting condition of Session 1 are introduced to regression analysis in the survey. As a result of variable selection optimisation, the variable of pupillary oscillation "s_ff2" was selected for $1as$ model. Also, participant's age contributes to regression analysis consistently. In general, pupil size decreased with age, as age-related miosis and cataracts may have also influenced PLR. As the factor of age may be the key for detecting the level of dementia, a more detailed investigation should be considered rather than stating the actual patient age, such as sorting age into groups using gradients.

In order to represent appropriate features of PLR waveforms, two sets of binocular PLR features, which respond to a light pulse to either eye, are introduced for analysis, such as (R+L) and (R-L) or independent variations of these, to the right and left eyes. As a result of variable selection processing, differential features such as "brd_f3" and an interfering feature such as "brl_f3" were selected as variables which were significant. These features are based on blue light pulses. In this study, it was hypothesised that dementia might negatively influence the synaptic signal transfer system of the oculo-motor nerve and the function of the pretectum. The above results may reflect this kind of impairment.

All features of PLR waveforms were extracted from a short constriction phase of about 3 out of 10 seconds of observation. The restoration phase of PLR may have significant features for detection of dementia since cross-correlation coefficients between the two eyes are reduced during the period. The measurement equipment was partially modified, in order to perform measurements at 60 Hz and introduce light sources using two wavelengths. Though each PLR waveform was confirmed manually, the measurement accuracy should be evaluated to permit a wider introduction of this system.

More significant feature detection and more robust procedures for prediction of the level of dementia should be developed. These topics will be the subject of our further study.

4.6 Regression Prediction Improvement Using Addition Information from Either Eye

In order to improve prediction performance of the level of dementia, additional information about PLR reactions is introduced. Also, observed PLR metrics are divided into separate measurements for the left and right eyes, and an additional prediction procedure is discussed in the sections that follow.

4.6.1 Ocular Oscillation During PLR

Additional features of ocular responses during PLRs are calculated in consideration of other phenomena such as post-illumination pupil response [58].

- Nystagmus
  Rapid involuntary eye movement is evaluated. The equipment used cannot observe eye movement, but the centre position of the pupil is used as an index of viewing direction. The respective frequency powers of the horizontal, vertical and cross components over frequency range of ~30 Hz are calculated. In order to compress dimensions, two factor scores are extracted from the frequency powers as 48 variables.

- Pupillary oscillation
  Frequency powers of pupillary oscillation during PLRs and when pupils were in a rest condition were extracted in Session 1. The three mean frequency powers are calculated as 24 variables.

4.6.2 Relationship Between Responses of Irradiated and Non-Irradiated Eyes

As indicated in the procedure above, PLRs of each eye were measured using irradiated light pulses. In order to divide the observations into sets for the irradiation of the left and right eyes, two merging indices such as "product" and divided "rate" of factor scores for each eye were introduced, as shown in the following equations [40]:

$$Product_{\{b|r,F_j\}} = iFactor_{b|r}j \times xFactor_{b|r}j, (j = 1, ., 3)$$

$$Rate_{\{b|r,F_j\}} = \frac{iFactor_{b|r}j}{xFactor_{b|r}j}, (j = 1, ., 3)$$

In the equations, $iFactor_{b|r}j$ means jth ($j = 1, 2, 3$), for factor scores of eyes irradiated using blue or red light pulses. $xFactor_{b|r}j$ means jth for factor scores of eyes not irradiated by blue or red light pulses. The relationships of these scores for each group of participants are shown in Figs. 4.15, 4.16, and 4.17. The distributions between three groups are slightly different, and the metrics may contribute to the classification.

Additional metrics for nystagmus and pupillary oscillation were also applied to the above calculations in order to evaluate left and right irradiated eye data sets separately.

4.6.3 Results of Prediction Performance

The possibility of identifying patients with AD or MCI has been examined using two logistic regression functions and the feature sets of PLRs. The two functions consist of means of detecting MCI+AD and AD patients, while prediction accuracy is evaluated using an AUC (area under the curve) metric for binary classification. The contribution of these features of PLR characteristics to evaluate the probability of accurate patient diagnosis will be discussed in greater detail.

Fig. 4.15 Relationship between product and rate of factor scores [AD]

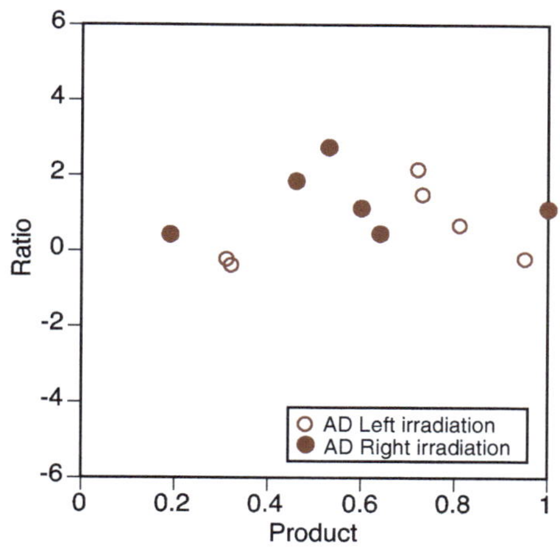

Fig. 4.16 Relationship between product and rate of factor scores [MCI]

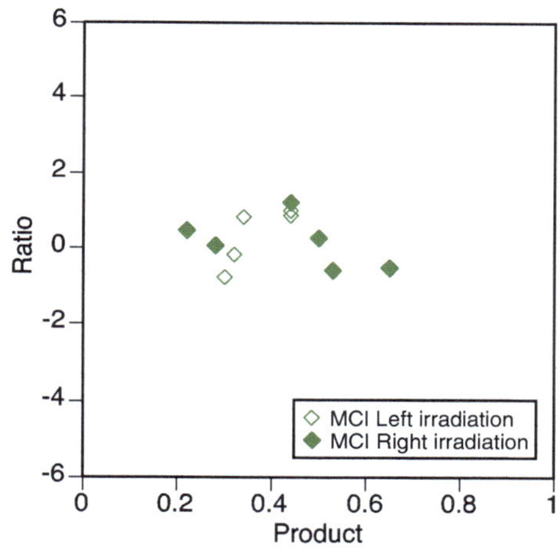

4.6.3.1 Features of Ocular Oscillation

Additional features of participant groups are compared in order to evaluate their contribution to classification. Some factor scores for frequency powers in groups of participants increased. There are some significant differences in factor scores of vertical oscillation for blue light pulses between NC and MCI+AD groups, but there are no significant differences between MCI and AD groups. For pupillary oscillation, all mean frequency powers in MCI and AD groups decrease. There are significant differences between the three groups, and most means (MCI+AD) are significantly lower than the means of the NC group ($p < 0.01$). The results may depend on a mechanism common during pupillary oscillations without light pulses in Condition 1.

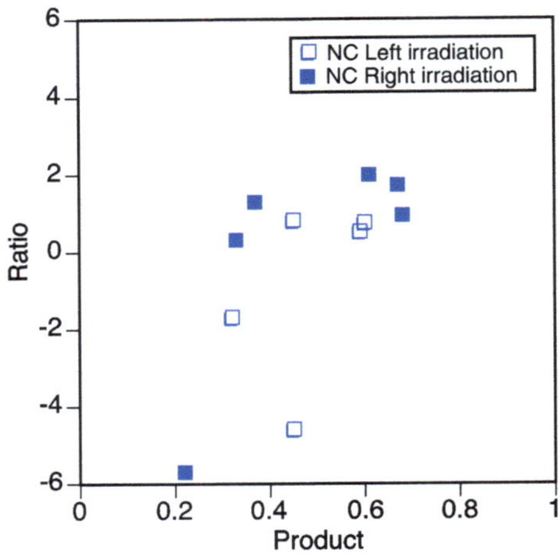

Fig. 4.17 Relationship between product and rate of factor scores [NC]

In summary, participants with cognitive performance problems have low stability of eye position and low pupillary oscillation. The phenomenon of senile miosis is well known in elderly people, and it may influence the measurements.

4.6.3.2 Classification Performance Using Extracted Features

First, the contribution of participant's age on prediction performance is examined. The results for MCI+AD and AD groups are shown in Table 4.8. Both AUC values are over 0.6, and once again age is confirmed as the most contributing variable.

Prediction performance is evaluated using AUC and AUC with cross-validation and variable selection using a step-wise procedure. The sixth model in Table 4.8 is formed using selected variables of PLR waveform features, and the use of these selected variables has been reported in a previous study [40]. In general, CV prediction performance is lower than ordinary prediction performance, and the difference shows the instability of the trained data.

The selected variables indicate features in response to blue light pulses such as "IbF1-3", "XbF2-3" etc. Another set of models (6**B) is focused on features in response to blue light pulses. However, both AUC and CV-AUC for the set of blue lights are comparable or less than the metrics with blue and red light pulses.

4.6.3.3 Classification Performance Using Ocular Oscillation

A possible classification model using all features was considered, but the number of variables was too great for a regression function, however. Some procedures to extract the most significant variables were tried in advance, such as the Boruta technique [29]. As a result of trial and error testing, a selection of common variables

Table 4.8 Logistic regression models for MCI and AD with NC groups

Model	Variables of extracted features	No. of variables	SEL	AUC	CV-AUC
MCI	age	1	–	0.72	0.70
AD	age	1	–	0.62	0.57
6nLs*	age,IbF1-3,IrF2,XbF2-3,PbF2,PrF2,DbF2,DrF3,PSD1-3	28	13	0.87	**0.76**
6nRs*	age,IbF1-3,IrF2,XbF2-3,PbF2,PrF2,DbF2,DrF3,PSD1-3	28	13	0.88	0.75
6aLs*	age,IbF2-3,XbF2-3,IrF3,XrF1,DbF1,PrF1-3,PSD1-2	28	13	0.83	0.66
6aRs*	age,IbF2-3,XbF2-3,IrF3,XrF1,DbF1,PrF1-3,PSD1-2	28	13	0.83	0.53
6nLB	age,XbF2-XbF3,DbF1-DbF2,PSD1-3,IbF1-IbF3,PbF2	16	12	0.85	0.72
6nRB	age,XbF2-XbF3,DbF1-DbF2,PSD1-3,IbF1-IbF3,PbF2	16	12	0.87	0.77
6aLB	age,IbmF1-IbmF3,XbmF3,PSD1	16	6	0.77	0.65
6aRB	age,IbmF1-IbmF3,XbmF3,PSD1	16	6	0.78	0.77
9nLs	age,XbF3,IbF3,DbF2,DrF3,PSD2 + IbyF1,PryF2,DbyF2,DbxF1,XrxyF2,IbpF2	100	12	0.88	**0.76**
9nRs	age,XbF3,IbF3,DbF2,DrF3,PSD2 + IbyF1,PryF2,DbyF2,DbxF1,XrxyF2,IbpF2	100	12	0.92	**0.79**
9nLB	age,IbF3,XbF1,SbF1-2,DbF2,PSD2 + IbyF1-2,PbpF3,DbxF2,IbxyF1,DbyF1,PbyF1,Xbxf1	52	16	0.90	0.72
9nRB	age,IbF3,XbF1,SbF1-2,DbF2,PSD2 + IbyF1-2,PbpF3,DbxF2,IbxyF1,DbyF1,PbyF1,Xbxf1	52	16	0.89	0.70
9aLs	IbF2-IbF3,PrF2 + XrpF2-XrpF3 **	100	5	0.80	**0.73**
9aRs	age,DrF3 + XbpF2,DrxyF2,XrpF3,PrpF1,PbpF3,PrpF3,IbxyF1 **	100	9	0.91	**0.83**
9aLB	age,IbF3,PbF2,PbF3,IbF2,PSD2 + DbxF2,PbxyF2,DbxF1,IbyF2,DbxyF1	52	11	0.88	0.65
9aRB	age,IbF3,PbF2,PbF3,IbF2,PSD2 + DbxF2,PbxyF2,DbxF1,IbyF2,DbxyF1	52	11	0.82	0.54

*: reported previously [40];
**: variable selection for left and right eyes was conducted separately

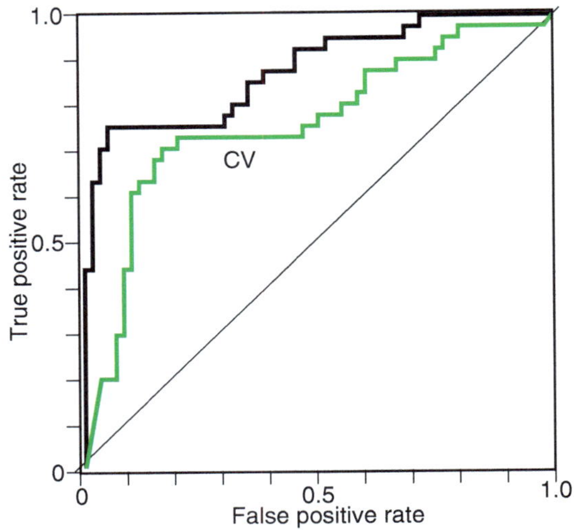

Fig. 4.18 AUC results for model "9nLs" with CV results (AUC=0.88, CV-AUC=0.76)

such as "9nL" and "9nR" was made from the sets of observation data for both the left and right eyes in order to classify NC vs. MCI+AD. The models consist of five variables from PLR waveform features, six variables from ocular oscillation features and age. The ROC curve for "9nLs" is shown in Fig. 4.18. Both the AUC and the CV-AUC for the updated model increase slightly from the results using the previous models. This result confirms the effectiveness of introducing ocular oscillation features.

However, variables common to the set of light pulses for the left and right eyes could not be selected in order to extract patients with AD during the variable selection procedure. Therefore, two functions for the left and right data sets, such as "9aLs" and "9aRs", are created independently. Both functions employ some variables from features of ocular oscillation. Classification performance improved since the functions themselves have improved.

The estimated probabilities of patients are summarised in Fig. 4.19 for the left eye data set and Fig. 4.20 for the right eye data set. The horizontal axis represents the probability of MCI or AD, and the vertical axis represents the probability of AD in both figures. As both of these figures show, most marks indicating NC participants are located in a corner area on both axes with a probability of less than 0.5. Values for MCI and AD participants are distributed in a wide range of probabilities. The two figures show the results from the data extracted for the left and right eye light pulses independently, using marks to indicate the plots for the left and right eyes of all participants in each of the two corresponding figures.

Fig. 4.19 Prediction performance probabilities for MCI+AD and AD using irradiated left eye data

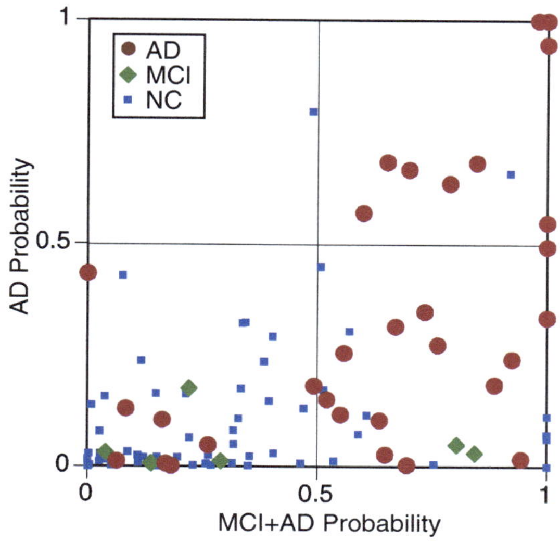

Fig. 4.20 Prediction performance probabilities for MCI+AD and AD using irradiated right eye data

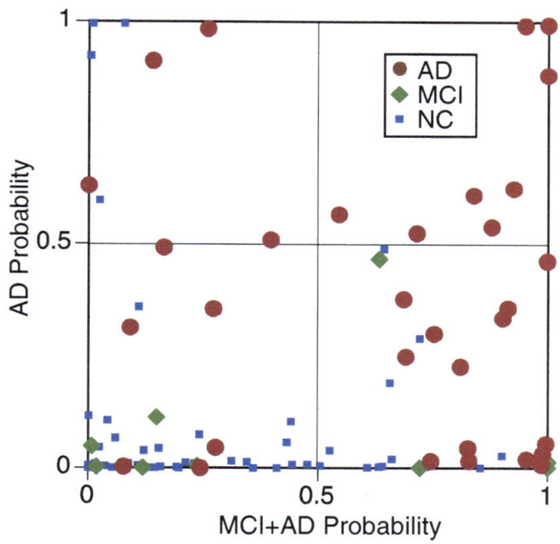

4.6.4 Discussion

First, two chromatic light pulses are required for accurate classification in a comparison of the performance of each eye. This result may support the conventional research hypothesis that PLRs based on different types of retinal ganglion cells should be compared to detect any impairment in signal transfer via optic nerves. In this approach, the effect of light intensity on PIPR waveforms should be considered [58]. Additional experimental protocols should also be considered.

For the classification of AD patients, the number of AD patients who can participate in the experiment is limited, and some of the features of responses for the left and right eyes are different. Therefore, a set of common variables of observed features cannot be selected for the classification of AD patients using left and right eye data. This problem of asymmetry may be due to the number of AD participants or the impairment of one or both eyes. If this major issue with the asymmetry of the results is shown to be dependent on the type of impairment, further improvements to our research procedure will be required.

In order to separate patients into groups accurately, as many features likely indicating their types of disease as possible were gathered for this work. However, if the selection of variables is correlated with each other, this issue may be an obstacle to the selection of variables which contribute to the classification of participants. When considering the variables to be selected, other approaches to extracting features should be taken into account and evaluated.

The two data sets of left and right light eye pulse trial data were originally created to reduce the repetition of measurements [40]. However, these two sets of data produce independent classification results. In comparing the two sets of results, the rate of misclassification of patients can be reduced. Since there are many observed metrics, sub-level classification using an ensemble technique may be possible.

Oscillation features were extracted from 10 seconds of recorded observations. As it has been suggested that PLRs consist of several phases such as transient constriction, sustained constriction and restoration to the baseline [24], these features should be extracted chronologically.

Careful consideration of these issues will be subjects of our further study.

4.7 Conclusion

Procedures for predicting the level of cognitive impairment of senior citizens were prepared using the binocular pupil light reflex (PLR) for chromatic light pulses, other metrics of pupillary changes and portable measuring equipment. Features of oculo-motors during PLR reactions generated using blue and red light pulses were extracted and compared. Participants were classified into three levels of dementia using logistic regression analysis of these features.

The results obtained are summarised as follows:

1. PLRs of the left and right eyes when light pulses are directed at either eye were compared. In addition, the latent factor scores of PLR features were also extracted. There are no significant differences in features and factor scores between the left and right eyes, however.
2. Factor scores and participant's ages were analysed in order to classify individuals into groups such as AD+MCI and NC. Participant's age information contributed to the classification of the groups. During regression analysis using a variable

selection procedure, factor scores for blue light pulses were extracted. PLRs for blue light pulses are key to accurate prediction.

3. Some cross-responding waveform features of PLRs of both eyes were selected as significant variables of regression functions in order to calculate the probability of the level of dementia. This validated the possibility that effective features of PLRs have an influence on the light reflex pathway when it is impaired by dementia.
4. PLRs between the left and right eyes are compared when light pulses are directed at either eye. In addition, the latent factor scores of features of PLRs were also extracted. There are no significant differences in features and factor scores between the left and right eyes, however.
5. Factor scores and participant's ages were analysed in order to classify individuals into groups such as AD+MCI and NC. Participant's age information contributed to the classification of the groups. During regression analysis using a variable selection procedure, factor scores for blue light pulses were extracted. PLRs for blue light pulses are key to accurate prediction.

The results suggest that the significance of a carefully conducted experimental procedure using a series of light conditions for PLR for diagnosis of AD and MCI and various features including the assessment of the left and right eyes contribute to classification performance. However, prediction accuracy of cognitive functions is still insufficient for clinical use. More clear evidence from the measurement of ocular reactions should be documented. In addition, a more robust procedure for feature extraction and prediction may require a more detailed set of features that are based on the mechanisms of PLR reactions.

The extraction and examination of these mechanisms will be subjects of our further study.

Acknowledgments This research was partially supported by the Japan Science and Technology Agency (JST), Adaptable and Seamless Technology transfer programme through target driven R&D (A-STEP) [JPMJTM20CQ, 2020–2022].

The authors would like to thank Prof. Masatoshi Takeda and Prof. Takenori Komatsu of Osaka Kawasaki Rehabilitation University, Toshinobu Takeda, MD at the Jinmeikai Clinic, Yasuhiro Ohta and Takato Uratani of the Uratani Lab Company Ltd. for their kind contributions.

References

1. Aoki, K., Ngo, T.T., Mitsugami, I., Okura, F., Niwa, M., Makihara, Y., Yagi, Y., Kazui, H.: Early detection of lower MMSE scores in elderly based on dual-task gait. IEEE Access **7**, 40085–40094 (2019)
2. Arevalo-Rodriguez, I., Segura, O., Solà, I., Bonfill, X., Sanchez, E., Alonso-Coello, P.: Diagonostic tools for alzheimer's disease dementia and other dementias: an overview of diagnostic test accuracy (DTA) systematic reviews. BMC Neurol. **14**(183), 1–8 (2014)
3. Asanad, S., Ross-Cisneros, F.N., Barron, E., Nassisi, M., Sultan, W., Karanjia, R., Sadun, A.A.: The retinal choroid as an oculavascular biomarker for Alzheimer's dementia: A histopathological study in severe disease. Alzheimer's & Dimentia: Diagnosis, Assessment & Diesease Monitoring **11**, 775–783 (2019)

4. Beatty, J.: Task-evoked pupillary response, processing load, and the structure of processing resources. Psychol. Bull. **91**(2), 276–292 (1982)
5. Bittner, D.M., Wieseler, I., Wilhelm, H., Riepe, M.W., Müller, N.G.: Repetitive pupil light reflex: potential marker in Alzheimer's disease? J. Alzheimers Dis. **42**, 1469–1477 (2014)
6. Breijyeh, Z., Karaman, R.: Comprehensive review on Alzheimer's disease: causes and treatment. Molecules **25**(5789), 1–28 (2020)
7. Brozou, C.G., Fotiou, D., Androudi, S., Theodoridou, E., Giantselidis, C., Alexandridis, A., Brazitikos, P.: Pupillometric characteristics in patients with choroidal neovascularisation due to age-related macular degeneration. Eur. J. Ophthalmol. **19**(2), 254–262 (2009)
8. Chaitanuwong, P., Singhanetr, P., Chainakul, M., Arjkongharn, N., Ruamviboonsuk, P., Grzybowski, A.: Potential ocular biomarkers for early detection of Alzheimer's disease and their roles in artificial intelligence studies. Neurol. Ther. **12**, 1517–1532 (2023)
9. Chávez-Fumagalli, M.A., Shrivastava, P., Aguilar-Pineda, J.A., Nieto-Montesinos, R., Del-Carpio, G.D., Peralta-Mestas, A., Caracela-Zeballos, C., Valdez-Lazo, G., Fernandez-Macedo, V., Pino-Figueroa, A., Vera-Lopez, K.J., Cardenas, C.L.L.: Diagnosis of Alzheimer's disease in developed and developing countries: systematic review and meta-analysis of diagnostic test accuracy. J. Alzheimer's Dis. Rep. **5**, 15–30 (2021)
10. Chougule, P.S., Najjar, R.P., Finkelstein, M.T., Kandiah, N., Milea, D.: Light-induced pupillary responses in Alzheimer's disease. Front. Neurol. **10**(360), 1–12 (2019)
11. Feigl, B., Zele, A.J.: Melanopsin-expressing intrinsically photosensitive retinal ganglion cells in retinal disease. Ophthalmol. Vis. Sci. **91**(8), 894–903 (2014)
12. Ferrari, C., Sorbi, S.: The comlexity of Alzheimer's disease: an evolving puzzle. Physiol. Rev. **101**, 1047–1081 (2021)
13. Folstein, M.F., Folstein, S.E., McHugh, P.R.: MINI-MENTAL STATE—a practical method for grading the cognitive state of patients for the clinician. J. Psychiatr. Res. **12**, 189–198 (1975)
14. Fotiou, D., Kaltsatou, A., Tsiptsios, D., a Nakou, M.: Evaluation of the cholinergic hypothesis in Alzheimer's disease with neuropsychological methods. Aging Clin. Exp. Res. **27**, 727–733 (2015). https://doi.org/10.1007/s40520-015-0321-8
15. Fotiou, D.F., Setergiou, V., Tsiptsios, D., Lithari, C., Nakou, M., Karlovasitou, A.: Cholinergic deficiency in Alzheimer's and Parkinson's disease: evaluation with pupillometry. Int. J. Psychophysiol. **73**, 143–149 (2009)
16. Frost, S., Kanagasingam, Y., Sohrabi, H., Bourgeat, P., Villemagne, V., Rowe, C.C., Macaulay, L.S., Szoeke, C., Ellis, K.A., Ames, D., Masters, C.L., Rainey-Smith, S., Martins, R.N., Group, A.R.: Pupil response biomarkers for early detection and monitoring of Alzheimer's disease. Curr. Alzheimer Res. **10**(9), 931–939 (2013)
17. Gamlin, P.D., McDougal, D.H., Pokorny, J.: Human and macaque pupil responses driven by melanopisn-containing retinal ganglion cells. Vis. Res. **47**, 946–954 (2007)
18. Granholm, E.L., Panizzon, M.S., Elman, J.A., Jak, A.J., Hauger, R.L., Bondi, M.W., Lyons, M.J., Franz, C.E., Kremen, W.S.: Pupillary responses as a biomarker of early risk for Alzheimer's disease. J. Alzheimer's Dis. **56**(4), 1419–1428 (2017)
19. Haj, M.E., Chapelet, G., Moustafa, A.A., Boutoleau-Bretonnière, C.: Pupil size as an indicator of cognitive activity in mild Alzheimer's disease. EXCLI J. **21**, 307–316 (2022)
20. Jarrold, W., Peintner, B., wilkins, D., Vergryi, D., Richey, C., Gorno-Tempini, M.L., Ogar, J.: Aided diagnosis of dimentia type through computer-based analysis of spontaneous speech. In: Proceedings of Workshop on Computational Linguistics and Clinical Psychology: From Linguistic Signal to Clinical Reality, pp. 27–37. Association for Computational Linguistics, New York (2014)
21. Kankipati, L., Girkin, C.A., Gamlin, P.D.: Post-illumination pupil response in subjects in without ocular disease. Invest. Ophthalmol. Vis. Sci. **51**(5), 2764–2769 (2010)
22. Kankipati, L., Girkin, C.A., Gamlin, P.D.: The post-illumination pupil response is reduced in Glaucoma patients. Vis. Neurosci. **52**(5), 2287–2292 (2011)
23. Kardon, R.H., Anderson, S.C., Damarjian, T.G., Grace, E.M., Stone, E., Kawasaki, A.: Chromatic pupillometry in patients with retinitis Pigmentosa. Ophthalmology **118**(2), 376–381 (2011)

24. Kawasaki, A., Kardon, R.H.: Intrinsically photosensitive retinal ganglion cells. J. Neuroophthalmol. **27**, 195–204 (2007)
25. Kelbsch, C., Strasser, T., Chen, Y., Feigl, B., Gamlin, P.D., Kardon, R., Peters, T., Roeckeln, K.A., Steinhauer, S.R., Szabadi, E., Zele, A.J., Wilhelm, H., Wilhelm, B.J.: Standards in pupillography. Front. Neurol. **10**(129), 1–26 (2019)
26. Khoury, R., Ghossoub, E.: Diagnostic biomarkers of Alzheimer's disease: a state-of-the-art review. Biomarkers in Neuropsychiatry **1**, 1–6 (2019)
27. Klumpp, P., Fritsch, J., Noeth, E.: ANN-based Alzheimer's disease classification from bag of words. In: Proceedings of Speech Communication; 13th ITG-Symposium, pp. 1–4. IEEE, New York (2018)
28. Kuhlmann, J., Böttcher, M. (eds.): Pupillography: Principles, Methods and Applications. W. Zuckschwerdt Verlag, Munchen, Germany (1999)
29. Kursa, M.B., Rudnicki, W.R.: Feature selection with the boruta package. J. Stat. Softw. **36**(11), 1–13 (2010)
30. Lim, J.K.H., Li, Q.X., He, Z., Vingrys, A.J., Wong, V.H., Currier, N., Mullen, J., Bul, B.V., Nguyen, C.T.O.: The eye as a biomarker for Alzheimer's disease. Front. Neurol. **10**(536), 1–14 (2016)
31. Liu, Z., Yang, Z., Gu, Y., Liu, H., Wang, P.: The effeciveness of eye tracking in the diagnosis of cognitive disorder: a systematic review and meta-analysis. PLoS One **16**(7), e0254059 (2021)
32. Marcucci, V., Kleiman, J.: Biomarkers and their implications in Alzheimer's disease: a literature review. Explorary Res. Hypothesis Med. **6**(4), 164–176 (2021)
33. McDougal, D.H., Gamlin, P.D.: Autonomic control of the eye. Compr. Physiol. **5**(1), 439–473 (2015)
34. Molitor, R.J., Ko, P.C., Ally, B.A.: Eye movements in Alzheimer's disease. J. Altzheimers Dis. **44**(1), 1–12 (2015)
35. Morgia, C.L., Ross-Cisneros, F.N., Hannibal, J., Montagna, P., Sadun, A.A.: Melanopsin-expressing retinal ganglion cells: implications for human diseases. Vis. Res. **51**, 296–302 (2011)
36. Nakayama, M., Nowak, W., Ishikawa, H., Asakawa, K., Ichibe, Y.: Discovering irregular pupil light responses to chromatic stimuli using waveform shapes of pupillograms. EURASIP J. Bioinf. Syst. Biol. **#18**, 1–14 (2014)
37. Nakayama, M., Nowak, W., Zarowska, A.: Detecting symptoms of dementia in elderly persons using features of pupil light reflex. In: Proceedings of the Federated Conference on Computer Science and Information Systems (FedCSIS), pp. 745–749 (2022)
38. Nakayama, M., Nowak, W., Zarowska, A.: Prediction procedure for dementia levels based on waveform features of binocular pupil light reflex. In: Proceedings of ACM Eye-Tracking Research & Applications (ETRA), pp. 1–6 (2023)
39. Nakayama, M., Nowak, W., Zarowska, A.: Symptoms of dementia in elderly persons using waveform features of pupil light reflex. In: E. Ziemba, W. Chmielarz, J. Wątróbski (eds.) Information Technology for Management: Approaches to Improving Business and Society. FedCSIS-AIST 2022, ISM 2022, pp. 91–107. Springer, Cham (2023)
40. Nakayama, M., Nowak, W., Zarowska, A.: Using features of PLRs to chromatic light pulse irradiations of either eye to detect dementia in elderly persons. Commun. Papers 18th Conf. Comput. Sci. Intell. Syst. **37**, 201–207 (2023)
41. Nakayama, M., Nowak, W., Zarowska, A.: Extracting dementia symptoms in elderly using ocular motor activity during pupil light reflexes to chromatic light pulses on either eye. In: Proceedings of iV2024: 28th International Conference Information Visualisation, pp. 75–80 (2024)
42. Nie, J., Qiu, Q., Philips, M., Sun, L., Yan, F., Lin, X., Xiao, S., Li, X.: Early diagnosis of mild cognitive impairment based on eye movement parameters in an aging chinese population. Front. Aging Neurosci. **29**(Article 221), 1–11 (2020)
43. Nowak, W., Nakayama, M., Kręcicki, T., Hachoł, A.: Detection procedures for patients of Alzheimer's disease using waveform features of pupil light reflex in response to chromatic stimuli. EAI Endorsed Trans. Pervasive Health Technol. **6**, 1–11 (2020). E6

44. Nowak, W., Nakayama, M., Kręcicki, T., Trypka, E., Andrzejak, A., Hachoł, A.: Analysis for extracted features of pupil light reflex to chromatic stimuli in Alzheimer's patients. EAI Endorsed Trans. Pervasive Health Technol. **5**, 1–10 (2019). E4
45. Nowak, W., Nakayama, M., Trypka, E., Zarowska, A.: Classification of Alzheimer's disease patients using metric of oculo-motors. In: Proceedings of the Federated Conference on Computer Science and Information Systems (FedCSIS), pp. 403–407 (2021)
46. Oh, A.J., Amore, G., Sultan, W., Asanad, S., Park, J.C., Romagnoli, M., Morgia, C.L., Karanjia, R., Harrington, M.G., Sadun, A.A.: Pupillary evaluation of melanopsin retinal ganglion cell function and sleep-wake activity in pre-symptomatic Alzheimer's disease. PLoS One **14**(12), 1–17 (2019)
47. Opwonya, J., Doan, D.N.T., Kim, S.G., Kim, J.I., Ku, B., Kim, S., Park, S., Kim, J.U.: Saccadic eye movements in mild cognitive impairment and Alzheimer's disease: a systematic review and meta-analysis. Neuropsychol. Rev. **32**, 193–227 (2022)
48. Peysakhovich, V., Causse, M., Scannella, S., Dehais, F.: Frequency analysis of a task-evoked pupillary response: luminance-independent measure of mental effort. Int. J. Psychophysiol. **97**, 30–37 (2015)
49. Porsteinsson, A., Isaacson, R., Knox, S., Sabbagh, M., Rubino, I.: Diagnosis of early Alzheimer's disease: clinical practice in 2021. J. Prev Alzheimers Dis. **8**, 371–386 (2021)
50. Sabeti, F., Maddess, T., Essex, R.W., James, A.C.: Multifocal pupillographic assessment of age-related macular degeneration. Optom. Vis. Sci. **88**(12), 1477–1485 (2011)
51. van der Schaar, J., Visser, L.N.C., Bouwman, F.H., Ket, J.C.F., Scheltens, P., Bredenoord, A.L., van der Flier, W.M.: Considerations regarding a diagnosis of Alzheimer's disease before dementia: a systematic review. Alzheimer's Res. Ther. **14**(31), 1–12 (2022)
52. Scinto, L., Frosch, M., Wu, C., Daffner, K., Gedi, N., Geula, C.: Selective cell loss in Edinger-Westphal in asymptomatic elders and Alzheimer's patients. Neurobiol. Aging **22**(5), 729–736 (2001)
53. Tao, L., Wang, Q., Liu, D., Wang, J., Zhu, Z., Feng, L.: Eye tracking metrics to screen and assess cognitive impairment in patients with nuurlogical disorders. Neurol. Sci. **41**, 1697–1704 (2020)
54. Tombaugh, T., McDowell, I., Kristjansson, B., Hubley, A.: Mini-mental state examination (MMSE) and the modified MMSE (3MS): a psychometric comparison and normative data. Psychol. Assess. **8**(1), 48–59 (1996)
55. Turner, R.S., Stubbs, T., Davies, D.A., Albensi, B.C.: Potential new approaches for diagnosis of Alzheimer's disease and related dementia. Front. Neurol. **11**(496), 1–10 (2020)
56. Zele, A.J., Adhikari, P., Cao, D., Feigl, B.: Melanopsin and cone photoreceptor inputs to the afferent pupil light response. Front. Neurol. **10**(529), 1–9 (2019)
57. Zhou, L., Fraser, K.C., Rudzicz, F.: Speech recognition in Alzheimer's disease and in its assessment. In: Proceedings of INTERSPEECH 2016, pp. 1948–1952. ISCA, New York (2016)
58. Zivcevska, M., Blakeman, A., Lei, S., Goltz, H.C., Wong, A.M.F.: Binocular summation in postillumination pupil response driven by melanopsin-containing retinal ganglion cells. Vis. Neurosci. **59**, 4968–4977 (2018)

Chapter 5
Detecting Procedures for Dementia Symptoms Using Features in Differential Waveforms of the Pupil Light Reflex to Chromatic Stimuli

Abstract Procedures for predicting dementia levels were developed using features of waveform shapes of the pupillary light reflex (PLR) in response to chromatic light pulse stimuli. Features of waveform shapes of PLRs were extracted using a functional data analysis technique, and some irregular signal responses used to make the predictions were detected. The coefficient vectors of B-spline basis functions were introduced, and the number of basis functions was controlled to produce an optimised model. The prediction procedure for dementia levels was developed using the features extracted from PLRs during two types of observations. Results showed that some differential waveform shapes of PLRs are based on physiological characteristics contributed to the prediction.

5.1 Introduction

5.1.1 Waveform Analysis of the Pupil Light Reflex

Irregular signal detection is often applied to diagnostic procedures. Though the simple pupil reaction to a flash of light is well known as the pupil light reflex (PLR), the responses are influenced by conditions of retinal ganglion cells and optic nerves such as Edinger-Westphal nucleus [12]. In order to distinguish people pseudo

---

Chapter Co-authors: Minoru Nakayama, Wioletta Nowak, Tomasz Kręcicki, and Anna Żarowska
Originally published as follows:
Minoru Nakayama, Wioletta Nowak, Tomasz Kręcicki, "Detecting Alzheimer's Patients Using Features in Differential Waveforms of Pupil Light Reflex to Chromatic Stimuli", EAI Transactions on Intelligent Systems and Machine Learning Applications, pp. 1–7, July, 2024.
Minoru Nakayama, Wioletta Nowak, Anna Żarowska, "Development of a Procedure for Detecting Dementia Symptoms Using Features in Differential Waveforms of the Pupil Light Reflex", Proc. Biosignals 2025 (to be appeared).

M. Nakayama, W. Nowak, *Development of Diagnostic Procedures Using Features of Pupil Reactions*, Behaviormetrics: Quantitative Approaches to Human Behavior 21, https://doi.org/10.1007/978-981-95-4502-5_5

positive with aged-macular degeneration (AMD) or Alzheimer's disease (AD) from a normal control (NC) group, differential signals of pupil responses under several observation conditions are examined [2, 4, 5, 9, 14, 22]. In particular, pupil responses based on intrinsically photosensitive retinal ganglion cells (ipRGCs) [6] are often compared in order to detect irregular responses and used as diagnostic procedures with patients [8, 27]. Since PLR observation can be conducted in-person using a general-purpose device with a head-mounted display (HMD), the response data can be accumulated during people's daily lives. If specific features of PLR waveforms can be detected effectively using these observations, it may be possible to introduce an early detection procedure for Alzheimer's disease [16, 18, 24].

5.1.2 Problems of Feature Extraction from Waveform Shapes

The extraction of specific features may be the key to diagnosing patients. Though several trials have been conducted, as mentioned in previous works [5, 31], a more effective procedure should be considered, in particular, localised features in response to characteristics of PLRs to chromatic light. The differential responses caused by problems with the transfer of signals by the optic nerve of each eye may provide clinical information about the influence of dementia or Alzheimer's disease on the synchronisation of pupil reactions [3, 12, 13, 20]. Asynchronicity of PLRs of both eyes shows some symptoms of the disease in the optic nerve system [15], as some factors may influence PLRs, and most participants are elderly people. In order to observe the disparity between pupils, PLR measurements of both eyes using a light pulse to either eye may be a useful way of doing so [16, 18, 19]. Another issue is feature extraction of PLR waveforms. While the pupil constriction phase during PLR is focused on, other phases may provide significant information about symptoms of the disease.

Towards this aim, polynomial spline bases using a B-spline technique effectively extract local features of the functional data within the time series observations [28]. If some typical features of AD patients could be characterised from waveform shapes, an additional analytical procedure may be possible. As the features of AD patients show irregular responses, differences between AD and NC groups can be evaluated along with temporal changes in PLR waveforms. These analyses may provide new approaches to the diagnostic procedure.

In this chapter, a functional data analysis technique is introduced in order to extract all features of differential signals of PLRs and examine the possibility of predicting dementia levels using two sets of data of PLR observations of elderly participants.

5.2 Related Works

5.2.1 *Feature Extraction from Waveforms*

Major approaches are summarised in Table 5.1. Conventional feature extraction of signal waveform uses Discrete Fourier Transform (DFT), which consists of periodical functions. Features of waveforms or line drawn objects can be noted as Fourier Descriptors [25, 30] using DFT. Though a function-based approach can suppress some of the noise of observations, this technique is constructed using periodical signals without the capture of local features.

In order to emphasise local features, sampling specific feature points presents behavioural characteristics explicitly, as shown in line 2 of Table 5.1. This approach may focus on specific characteristics in waveform shapes. A feature points-based approach to detecting AD patients shows better performance than using features of Fourier descriptors, but performance remains at a limited level even when an Elastic-Net with Lasso technique is used [1], since local features without focused points may not be considered to be effective [22, 23]. Therefore, these analyses employed a set of fixed feature points which were extracted during a large reaction to the light pulse [22, 24].

An additional approach may be the use of functional data analysis by introducing polynomial spline bases such as B-spline basis [11, 15, 26], as shown at the bottom of Table 5.1. This method can be applied to waveforms flexibly, and it may be possible to detect feature contributions during temporal changes using regression analysis, though some tuning of the parameters would be required, however. The differences in performance of biological time series data between two bases functions such as Fourier and B-spline have been examined [28]. In considering the benefit of employing polynomial spline bases, the possibility of diagnosing AD patients using PLR waveforms and this technique is examined. Also, an appropriate dimension reduction technique will be required in order to suppress the large number of dimensions of the features.

In this chapter, functional data analysis using B-spline basis is simply referred to as functional data analysis (FDA) when described later.

Table 5.1 Comparing feature extraction from waveform shapes

Method	Basis function	Benefit/limitation
Fourier descriptor [25, 30]	Fourier basis	Periodic data analysis during observed duration, particularly useful for periodic data [14, 28]
Feature definition	n/a	Some specific features need to be defined in advance such as [23]. Shape similarity is implicitly required for the observation
Polynomial spline bases [10, 26, 28]	B-spline basis	Capturing local features of the functional data which contains local changes

5.2.2 *Diagnosis Using PLR Waveforms*

A number of clinical surveys measuring PLR reactions to chromatic light stimuli have been conducted since ipRGCs were discovered [4, 6, 8, 9]. PLR waveform shapes are influenced by the stimulus characteristics of light pulses, such as wavelength (light colour) and intensity (luminance), since the activation of irradiated ganglion cells in response to light stimuli is different. The distributions on the retina and the activated light wavelengths between ganglion cells, cones, rods and ipRGCs are different [5, 6]. As temporal changes in waveform shapes on ganglion cells are also different [31], the possibility of a diagnostic procedure using PLRs for chromatic light pulses has been suggested by several other researchers [2, 8, 27]. PLR waveforms are influenced by ageing and by problems with signal transfer pathways from the retina [3, 12, 13, 20]. Therefore, if the differential signals between the PLR waveforms could be analysed, these differential waveforms may reveal specific features about patients or features of their reaction mechanisms. In addition to these studies, possible levels of dementia have also been measured using chromatic light pulse stimuli of the pupil [13, 20]. After additional systematic surveys were conducted, further analytical reports were published [16, 18].

In order to create a discriminant function for AD patients, a sufficient number of PLR observations using both patients and a normal control group are required as a reference. With most clinical approaches, possible observation opportunities are limited, and individual differences are not insignificant, as well, however. Since the number of patients who can participate is limited, most analyses have been published as case studies. Therefore, an appropriate data extension technique needs to be applied to the features extracted from observational data sets using a non-parametric bootstrap technique [24], for example.

5.3 Pupil Light Reflex to Chromatic Light Pulses on Both Eyes

In this section, the contribution of differential signals of PLRs to the experimental process of detecting AD patients using a method of functional data analysis is examined.

The following topics are addressed:

1. The feature extraction procedure for differential signals between responses to chromatic stimuli or levels of light intensity is examined using a functional data analysis technique.
2. Classification performance for AD patients is evaluated using the summation and differential signals of PLRs and their features.

5.3.1 Data Set Measurements

5.3.1.1 PLR Observation

A single PLR waveform in response to a light pulse of aged people was measured [22]. The stimuli consisted of three chromatic lights, red (635 nm), blue (470 nm) and white (CIE x:0.28, y:0.31), at two levels of brightness (10 and 100 cd/m^2). The two factors, chromatic lights and light intensity, are keys for PLR waveforms. In total, six stimulus conditions were labelled as r10, r100, b10, b100, w10 and w100. The duration of observation was 10 seconds, with the first 2s being a pre-stimulus phase as a rest period, followed by a 1s light pulse and 7s as a restoration phase. Pupil diameters were measured in mm at 60 Hz using a system developed by some of the authors [21, 22]. PLRs for each stimulus were observed in single trials using a repeated-measure design.

A conventional PLR experiment was performed on 19 participants (42∼89 years old, mean age: 70.6), 12 of which were healthy individuals with normal vision (normal control (NC) group: 62∼89 years old, mean age: 72.1) and 7 of which were patients with Alzheimer's disease (AD patients: 42∼84 years old, mean age: 68.1) who had already been diagnosed by medical doctors. It was not easy to invite volunteers who were aged over 80 to participate in the experiment. Informed consent was obtained from all participants prior to the experiment.

Examples of measurements for a healthy individual and for an AD patient are shown in Figs. 5.1 and 5.2. In these figures, PLRs are illustrated in response to six stimuli, namely the three colours and two levels of brightness.

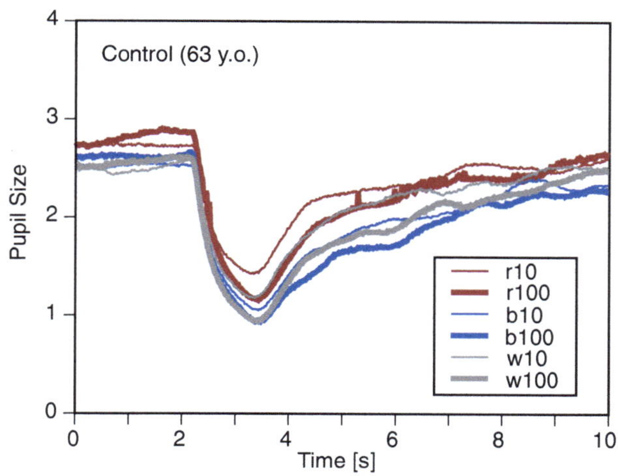

Fig. 5.1 An example of PLRs for a normal control subject

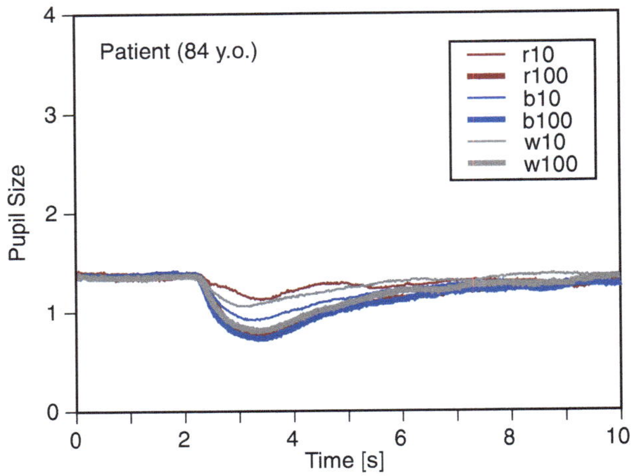

Fig. 5.2 An example of PLRs for an AD patient

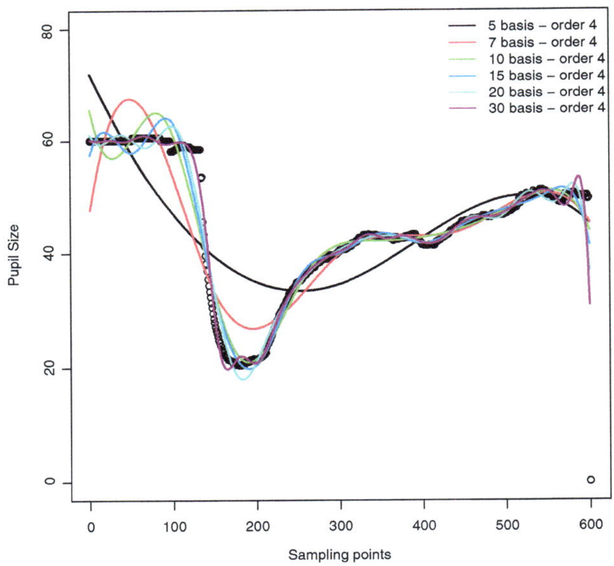

Fig. 5.3 A summation of PLRs for blue light pulses and B-spline representations using 5~30 basis functions

5.3.1.2 Functional Data Analysis

These waveforms were represented using B-spline basis functions as a technique of functional data analysis (FDA) technique [11, 26]. Figure 5.3 shows an example of an individual summation signal of PLRs for b10 and b100, which are blue

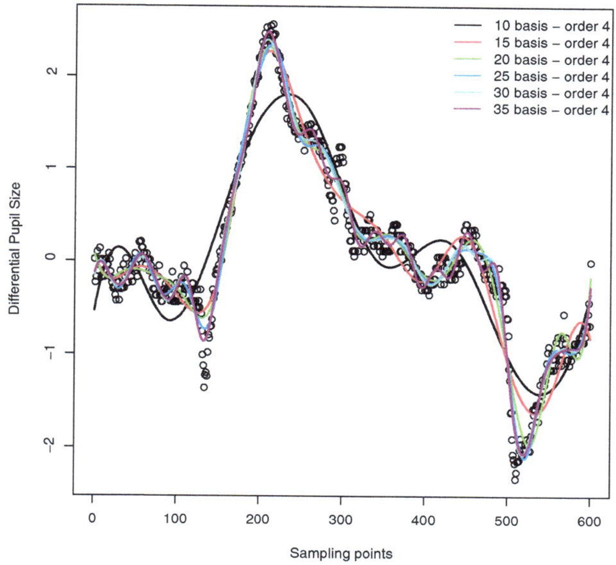

Fig. 5.4 A differential signal of PLRs for blue light pulses between 10 and 100 cd/m^2 and B-spline representations using 10∼35 basis functions

light pulses (the open circles). The summation signal seems moderated and with less noise. Some B-spline functions are introduced periodically, and the composite signal can represent the PLR waveform. When the number of functions increases, the fitness of the waveform increases, as shown in Fig. 5.3.

A differential signal of PLRs between b10 and b100 is illustrated in Fig. 5.4. As the figure shows however, there is little difference in advance of the light pulse, and the difference is maximised at around the most pupil-constricted time and in the final stage of pupil restoration. Since the differential signal shows spontaneous changes, noisy waveforms are presented in comparison with the summation signal. The FDA technique is also applied, as a greater number of functions are required to represent the waveform.

The coefficients for each B-spline function which represent the waveform can be a feature set of individual responses. Here, differences between chromatic stimuli or light intensities are processed, and feature sets of PLR waveforms are used for extracting patients with AD.

5.3.1.3 Data Extension and Discriminant Analysis

Since the experimental data consists of only 7 AD and 12 NC participants, the number of samples is insufficient to perform discriminant analysis. Here, a data extension technique is applied to the extracted feature set of sampled data using random Gaussian distribution numbers. As a result, 100 samples were generated for each AD and NC group from the statistics of the sampled data such as Gaussian data generation with means and SDs of each group of samples. This generated data set is used for training, and the original measured data is used for testing the trained models. Of course, the training data set and models can be re-generated to extend feature spaces when new experimental data is added. The data spaces of the extracted features of the two groups of participants were trained with a sparse logistic regression function using Elastic-Net with Lasso technique, a variable selection technique [1]. Because there is very little labelled data for detecting patients with AD, the probability of classification of each participant is calculated using data from their responses. Classification performance of the trained models was evaluated in regard to the number of basis functions using the sampled data set. This testing procedure may provide confirmation of the effectiveness of introducing features.

5.3.2 Results

Classification performance of the trained models using AD and NC participants is evaluated as a level of accuracy and as the area under the curve (AUC) of ROC for the sampled data.

5.3.2.1 Summation and Differentials Between Chromatic Stimuli

The effect of each chromatic stimulus is confirmed in this section. PLRs at the two levels of light intensity were observed, and summation signals for each chromatic light were calculated. These signal features were extracted using the FDA technique and were weighted for classification using Elastic-Net. The temporal changes of feature contributions as coefficient functions [10] for each chromatic stimulus for AD and NC were calculated, as shown in Fig. 5.5. The central dotted line indicates zero, meaning there is no effect on the classification. So, the drawing of a curved line far from zero may possibly be the key to classification. The main difference appeared at the beginning of observations because the pupil sizes of the two groups were different. Some differences appear during the PLR restoration process. In particular, the differences between blue and red light pulses are larger than the differences between white and blue or red light pulses.

Classification performance is summarised in Fig. 5.6. The features are generated along with the number of basis functions using FDA. As shown in Fig. 5.3, a small number of basis (n-basis) such as 10 or 20 are sufficient for the summation signal. For differential signals, many n-basis are required, as shown in Fig. 5.4. The horizontal axis in Fig. 5.6 represents the n-basis for the differential signals. The vertical axis represents the correct rates. The performances as percentage correct are sticked to some levels. When the feature vectors are combined, the number of dimensions for features increased. Therefore, performance has a peak

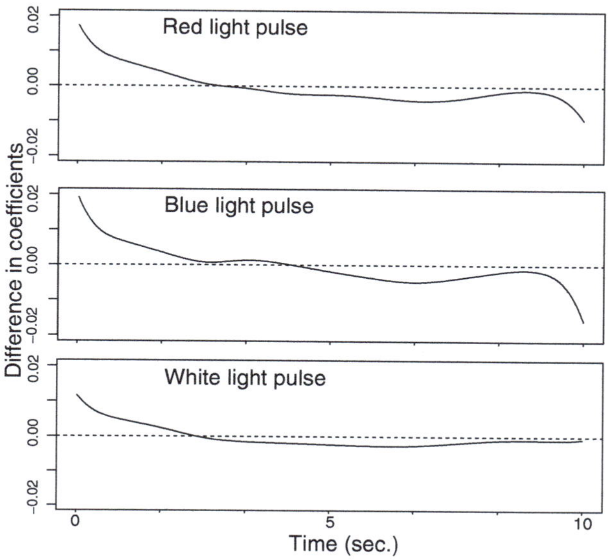

Fig. 5.5 Feature differences for chromatic stimuli with observation time for AD and NC subjects

Fig. 5.6 Classification performance for AD and NC participants using trained models, with summations (R-B-W) and differentials between chromatic stimuli

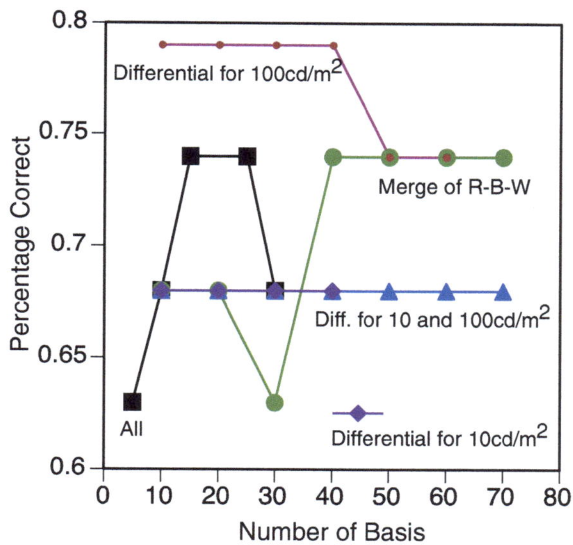

at around 10–20 n-basis when all generated features are introduced including summation signals (all, $AUC = 0.83$). For the summation signals, combinations of three chromatic stimuli (a merger of R-B-W) produce a plateau when the number of n-basis is between 40 and 70 ($AUC = 0.85$). However, peak performance comes from differentials between chromatic responses ($AUC = 0.88$) for high light intensities ($100\,cd/m^2$), though the performance of both low light intensity ($AUC = 0.80$) and a combination of two light intensities ($AUC = 0.82$) remains at around the middle level of performance. These results suggest the difference between the levels of light intensity may provide the information needed to diagnose patients with AD. Comparing the classification performance using a feature points-based approach provides an approximate value of the $AUC = 0.9$ [22], and the performance of most values for AUC is slightly lower. Nevertheless, the procedure may be applied to various conditions using summation and differential signals.

5.3.2.2 Summation and Differentials Between Stimulus Light Intensities

Differential signals between the two light intensities (10 and 100 cd/m^2) are examined, and the temporal changes in feature contributions between AD and NC participants for each chromatic condition are summarised using the same format as shown in Fig. 5.7. The differentials for red and blue light pulses may contribute to classification along with the duration of the observation time. Though the differentials for red and blue light pulses remained certain levels such as non-zero, differential level for white light pulses approached around zero. The tendency may show less contribution to discriminate performance.

Classification performance is summarised using the same format in Fig. 5.8. The highest performance is with red light, such as when there are more than 20 n-basis for differentials and 10 n-basis for summation of PLRs at the two intensities of light

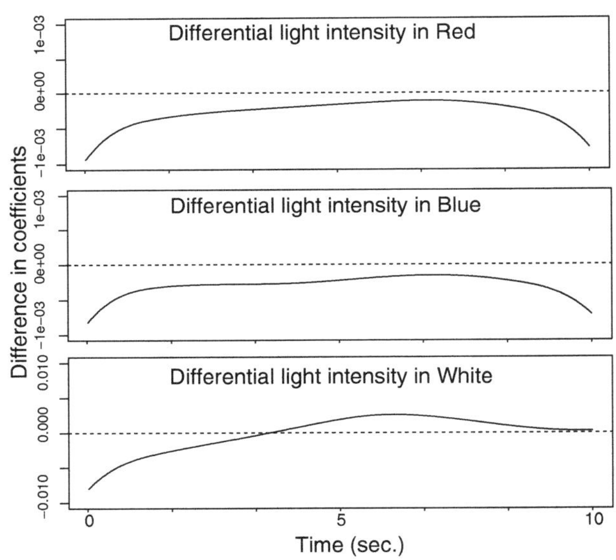

Fig. 5.7 Feature differences for differentials of light intensities with observation time of AD and NC subjects (differences for white light include a wider range than others)

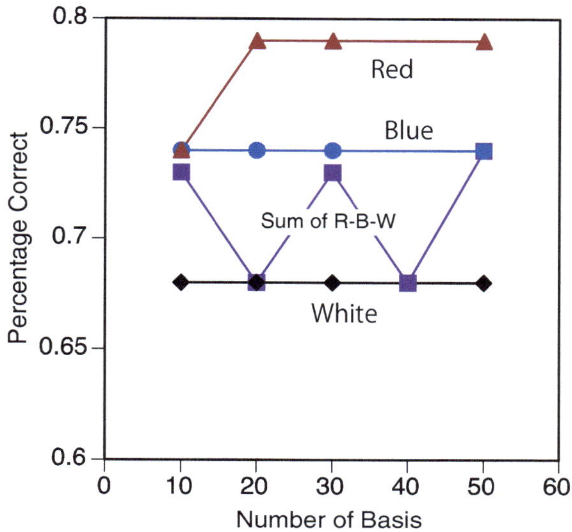

Fig. 5.8 Classification performance of AD and NC participants using trained models with summations and differentials between two light intensities

($AUC = 0.88$). Conditions using blue light show similar performance (percentage correct $= 0.74$, $AUC = 0.88$). Other conditions such as combinations including PLRs for white light remained at lower levels of performance.

These results suggest that PLRs for different light intensities or a high level of light intensity with red or blue lights may show better AD classification performance.

5.4 Analysing PLRs to Chromatic Light Pulses on Either Eye

This section examines the possibility of detecting dementia levels using differential features of waveform shapes of PLRs and a functional data analysis technique [15]. The following topics are addressed in this paper.

1. The features of waveform shapes, which are summations and ratios of PLRs of both eyes, are extracted using a B-spline technique, while eye is irradiated with either a blue or a red light pulse.
2. Classification performance of the level of dementia is examined using a combination of extracted features of PLRs.

5.4.1 Experimental Procedure

The set of data to be analysed was surveyed in a previous study [16, 18], which measured 1 sec chromatic light pulses (blue or red) to either eye during a 10 sec period of observation. The participants were elderly people and included some patients with Alzheimer's disease (AD), mild cognitive impairment (MCI), and a normal control group (NC).

5.4.1.1 Measurement Protocols

Pupil diameters of each eye were measured for 10 sec in a temporal darkened space under the following conditions [16].

1. Condition 1: Observe static pupil oscillation without light pulses
2. Condition 2: Blue light pulses to the right eye
3. Condition 3: Blue light pulses to the left eye
4. Condition 4: Red light pulses to the right eye
5. Condition 5: Red light pulses to the left eye

The light stimulus is irradiated for one second during each session. The combination of the experimental conditions is fixed for all participants. In regard to each of the four light pulse irradiation conditions, the measured data is classified into sets for irradiation of the left and right eyes. Condition 1 is a set of common data for both eyes [19].

Pupil diameters of both eyes were observed using a piece of specialised measuring equipment at 60 Hz (URATANI, HITOMIRU). Light stimuli consisted of a blue light source (469 nm, 14.3 cd/m^2, 6.5lx) and a red light source (625 nm, 12.3 cd/m^2, 10.5lx).

5.4.1.2 Experimental Participants

A total of 101 elderly individuals were selected to take part in the experiment (mean age: 78.5, SD: 8.9, F: 66, M: 35) [16, 17]. Their cognitive functions were diagnosed by a clinical doctor and sorted into three levels using MMSE (Mini-Mental State Examination) scores [29].

- AD: 32 (mean age: 83.0, SD: 6.3, F: 22, M: 10).
- MCI: 9 (mean age: 82.1, SD: 6.3, F: 5, M: 4).
- NC: 60 (mean age: 75.6, SD: 9.2, F: 38, M: 22).

The experimental procedure was approved by an ethics committee at Osaka Kawasaki Rehabilitation University.

5.4.2 *Analysis of Functional Data*

Measured pupil sizes are summarised into sets for irradiation of the left and right eyes. Condition 1 is provided as a common control condition. Observed signals

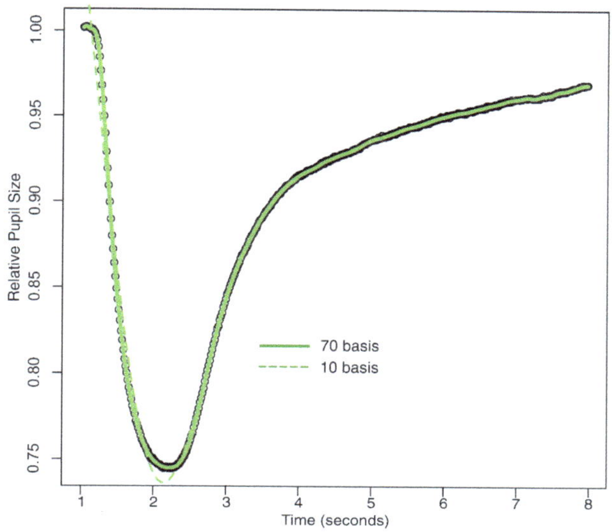

Fig. 5.9 A summation signal of PLRs for blue light pulses

from both eyes are processed as summation of the left and right eyes, and the ratios between the sizes for the left and right eyes (right size/left size) are asynchronous measurements. All data are standardised using the initial-level measurements made during the initial 1 sec before irradiation. Also, the period of the final 2 sec is eliminated in order to reduce the amount of data measured. Figure 5.9 shows the overall means of summation waveforms of the left and eye responses to right eye blue light irradiation. The horizontal axis represents the time course, from 1 to 8 seconds. A waveform of pupil size shows pupil constrict, with dome delay, after 1-second blue light pulse (1–2 sec). The summation waveform is very smooth. This waveform can be represented using a B-spline basis function as a functional data analysis technique [11, 26]. When 70 basis functions are introduced (the solid line), the fit is much better than in the case of ten basis functions (the dotted line), as shown in Fig. 5.9.

The individual differential ratio of both eyes is illustrated in Fig. 5.10. Most differential lines are connected around ratio = 1, though some differences are observed around light pulses (1–3 sec) and at the end of the observation period (8 sec). The red lines indicate patient cases (AD and MCI). A summation of these waveforms is illustrated in Fig. 5.11, showing that even the changes in overall summation have not been smoothened. In this case, 70 B-spline basis functions are required to represent the waveform.

As an FDA technique, sets of coefficients for basis functions can represent features of waveform shapes of PLRs or differential PLRs. These features are used for classification of participants in the section which follows.

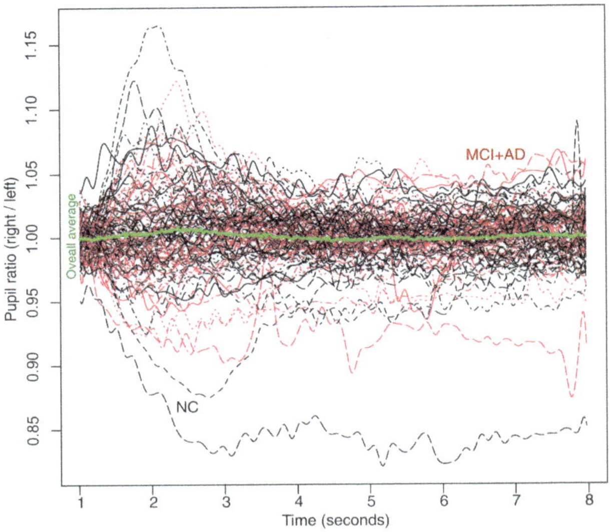

Fig. 5.10 Individual differential signals of PLRs for blue light pulses

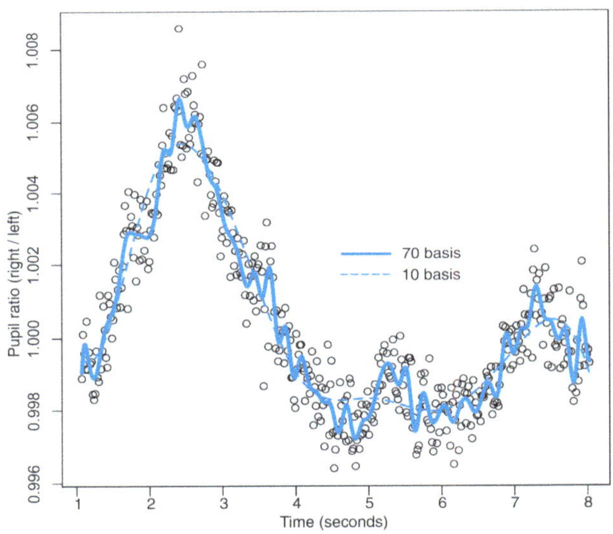

Fig. 5.11 A differential signal of PLRs for blue light pulses

5.4.3 Classification Results

Using the extracted features from waveform shapes and logistic regression functions, patients are classified as either patients (AD or MCI) or NC participants.

A logistic function presents the probability of diagnosing patients using the set of features mentioned above. The waveform features are generated from waveform shapes, and measurement conditions, such as using blue, red, or both colours of chromatic lights, can be combined. However, the number of dimensions of features increases even as the LASSO technique [7] is introduced in order to select features which are significant. During the feature selection process, information of participant's age was not selected. This procedure might be independent to an age factor.

In an evaluation of classification performance, cross-validation of the irradiation of the right or left eye is introduced, and another set of data of left and right eye measurements is used, for example, when the regression model is trained using the data set of the right eye and tested against the data set of the left eye. In comparison of performance using several combinations of features of waveforms, the best performance obtained to date is from a combination of using a control condition, and differential ratios for blue and red light, for a total of 270 dimensions of features.

A contingency table classifying performance is summarised in Table 5.2. The horizontal column represents the prediction results, and the vertical row shows the type of participant. Once again, the results present the classification of the data set of the left eye using a regression model trained with the data set of the right eye. In the results, 71 out of the 101 participants were classified correctly using a threshold 0.5, and the accuracy was 70%. In another regression model classification of the right eye, the accuracy was 65%. Though an improvement of prediction performance is required, the differential ratios of irradiations using both blue and red light pulses may indicate the existence of physiological impairment, which causes dementia.

Using the data set combination, two independent predictions are produced for each participant. These probabilities are summarised in Fig. 5.12. The horizontal axis represents the probability of patients trained with the right eye data set and tested using the left eye data set, and the vertical axis represents probabilities of patients trained with the left eye data set and tested using the right eye data set. In the figure, the blue dots represent NC participants, and the orange dots represent MCI or AD patients. The threshold of 0.5 is represented by fine lines along the horizontal and vertical axes. The plots of many participants overlap in the centre of the graph.

The two procedures produce two probabilities, so participants are classified again using the logical OR of the two prediction results. These results are summarised in Table 5.3. Using this procedure to merge the results, the number of patient

Table 5.2 Contingency table: right trained/left tested

	Predicted		
Participants	NC	MCI+AD	Total
NC	54	6	60
MCI+AD	24	17	41
Total	78	23	101

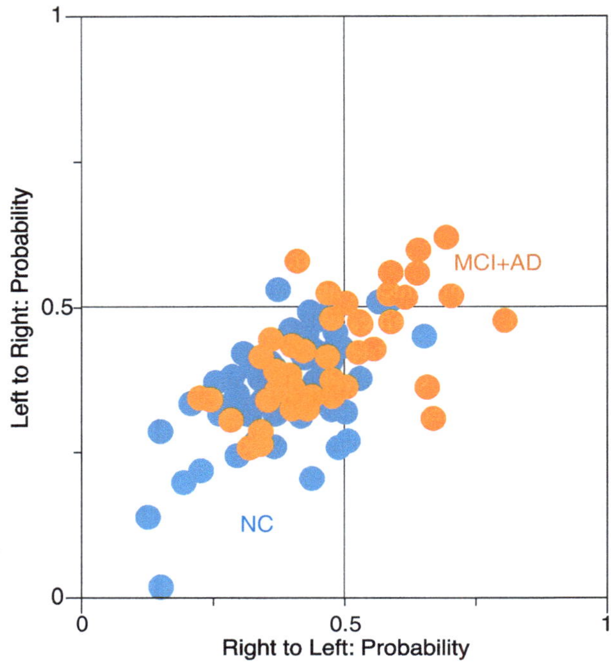

Fig. 5.12 A differential signal of PLRs for blue light pulses

Table 5.3 Contingency table using both predictions

	Predicted			
Participants	NC	Either(+)	Both(+)	Total
NC	52	6	2	60
MCI+AD	22	11	8	41
Total	74	17	10	101

predictions increases slightly. However, overall accuracy remains at the same level such as 70% in the case mentioned above.

In order to evaluate the contribution of extracted waveform features, the feature values are summarised, as shown in Fig. 5.13. The horizontal axis represents time such as a dimensional sequence, and the vertical axis represents the weight values. Mean weights of the three waveforms for control, blue, and red differentials of the data sets of the right and left eyes are summarised. The weights on both sides are almost the same. Regarding the changes in weights, the amounts are relatively large during the light pulses and at the ends of the observations. There are some differences between blue and red light pulses. Also, the weights for red light pulses are maintained after the light pulse. As a phenomenon, post-illumination pupil response (PIPR) is observed in the case of red light irradiation. These differences may suggest that some weight patterns at the ends of the reaction observations show some increase. These results indicate that the most correct predictions are accomplished by using the overall features of PLR waveform shapes.

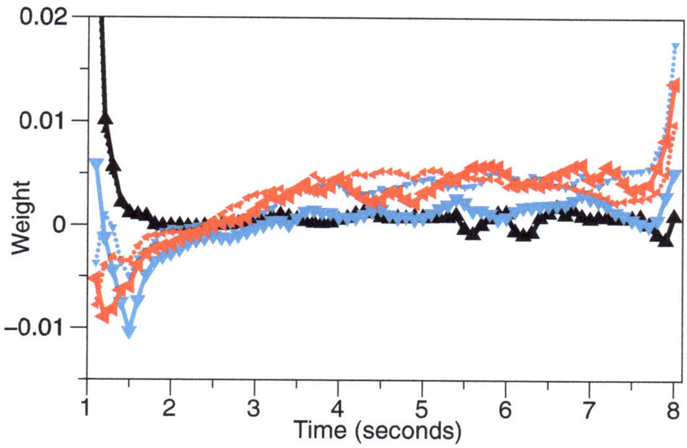

Fig. 5.13 Comparison of coefficients of PLRs for control (black), and differentials for blue and red light pulses

Further feature selection and additional features of waveforms from around the irradiation of light pulses when constriction of the pupil does not occur may be required in order to improve performance. A detailed analysis of performance improvements will be a subject of our further study.

5.5 Conclusion

This chapter presents procedures for predicting dementia levels using differential signals of PLRs in response to chromatic light stimuli under various conditions. The features of PLR waveforms were extracted using B-spline basis functions as an FDA technique. The number of basis functions was controlled in order to optimise classification performance. Classification was conducted using a trained Elastic-Net model. The features of waveform shapes may also represent physiological features of PLRs of chromatic light stimuli.

In this approach, all features of PLR waveforms can be extracted without difficulty, though an appropriate control for feature dimensions is required in order to improve prediction performance. This objective will be the subject of our further study.

References

1. Package 'glmnet'. https://cloud.r-project.org/web/packages/glmnet/glmnet.pdf
2. Asanad, S., Ross-Cisneros, F.N., Barron, E., Nassisi, M., Sultan, W., Karanjia, R., Sadun, A.A.: The retinal choroid as an oculavascular biomarker for Alzheimer's dementia: A

histopathological study in severe disease. Alzheimer's & Dimentia: Diagnosis, Assessment & Diesease Monitoring **11**, 775–783 (2019)
3. Chaitanuwong, P., Singhanetr, P., Chainakul, M., Arjkongharn, N., Ruamviboonsuk, P., Grzybowski, A.: Potential ocular biomarkers for early detection of Alzheimer's disease and their roles in artificial intelligence studies. Neurol. Ther. **12**, 1517–1532 (2023)
4. Chougule, P.S., Najjar, R.P., Finkelstein, M.T., Kandiah, N., Milea, D.: Light-induced pupillary responses in Alzheimer's disease. Front. Neurol. **10**(360), 1–12 (2019)
5. Fotiou, D.F., Setergiou, V., Tsiptsios, D., Lithari, C., Nakou, M., Karlovasitou, A.: Cholinergic deficiency in Alzheimer's and Parkinson's disease: evaluation with pupillometry. Int. J. Psychophysiol. **73**, 143–149 (2009)
6. Gamlin, P.D., McDougal, D.H., Pokorny, J.: Human and macaque pupil responses driven by melanopisn-containing retinal ganglion cells. Vis. Res. **47**, 946–954 (2007)
7. Kawano, S., Matsui, H., Hirose, K.: Statistical Modeling via Sparse Estimation. KYORITSU Publication, Tokyo (2018)
8. Kawasaki, A., Kardon, R.H.: Intrinsically photosensitive retinal ganglion cells. J. Neuroophthalmol. **27**, 195–204 (2007)
9. Kelbsch, C., Strasser, T., Chen, Y., Feigl, B., Gamlin, P.D., Kardon, R., Peters, T., Roeckeln, K.A., Steinhauer, S.R., Szabadi, E., Zele, A.J., Wilhelm, H., Wilhelm, B.J.: Standards in pupillography. Front. Neurol. **10**(129), 1–26 (2019)
10. Levitin, D.J., Nuzzo, R.L., Vines, B.W., J.O.R.: Introduction to functional data analysis. Can. Psychol. **48**(3), 135–155 (2007)
11. López, O.A.M., López, A.M., Crossa, J.: Multivariate Statistical Machine Learning Methods for Genomic Prediction. Springer, New York (2022)
12. McDougal, D.H., Gamlin, P.D.: Autonomic control of the eye. Compr. Physiol. **5**(1), 439–473 (2015)
13. Molitor, R.J., Ko, P.C., Ally, B.A.: Eye movements in Alzheimer's disease. J. Altzheimers Dis. **44**(1), 1–12 (2015)
14. Nakayama, M., Nowak, W., Ishikawa, H., Asakawa, K., Ichibe, Y.: Discovering irregular pupil light responses to chromatic stimuli using waveform shapes of pupillograms. EURASIP J. Bioinforma. Syst. Biol. **#18**, 1–14 (2014)
15. Nakayama, M., Nowak, W., Krecicki, T.: Detecting Alzheimers patients using features in differential waveforms of pupil light reflex for chromatic stimuli. EAI Endorsed Trans. Intell. Syst. Mach. Learn. Appl. **1**, 1–7 (2024)
16. Nakayama, M., Nowak, W., Zarowska, A.: Detecting symptoms of dementia in elderly persons using features of pupil light reflex. In: Proceedings of the Federated Conference on Computer Science and Information Systems (FedCSIS), pp. 745–749 (2022)
17. Nakayama, M., Nowak, W., Zarowska, A.: Prediction procedure for dementia levels based on waveform features of binocular pupil light reflex. In: Proceedings of ACM Eye-Tracking Research & Applications (ETRA), pp. 1–6 (2023)
18. Nakayama, M., Nowak, W., Zarowska, A.: Using features of PLRs to chromatic light pulse irradiations of either eye to detect dementia in elderly persons. In: Communication Papers of the 18th Conference on computer, Science and Intelligent Systems, vol. 37, pp. 201–207 (2023)
19. Nakayama, M., Nowak, W., Zarowska, A.: Extracting dementia symptoms in elderly using ocular motor activity during pupil light reflexes to chromatic light pulses on either eye. In: Proceedings of iV2024: 28th International Conference Information Visualisation, pp. 75–80 (2024)
20. Nie, J., Qiu, Q., Philips, M., Sun, L., Yan, F., Lin, X., Xiao, S., Li, X.: Early diagnosis of mild cognitive impairment based on eye movement parameters in an aging chinese population. Front. Aging Neurosci. **29**(Article 221), 1–11 (2020)
21. Nowak, W., Ząrowska, A., Szul-Pietrzak, E., Misiuk-Hojło, M.: System and measurement method for binocular pupillometry to study pupil size variability. BioMedical Eng. Online **13**(#69), 1–16 (2014)
22. Nowak, W., Nakayama, M., Kręcicki, T., Hachoł, A.: Detection procedures for patients of Alzheimer's disease using waveform features of pupil light reflex in response to chromatic stimuli. EAI Endorsed Transact. Pervasive Health Technol. **6**, 1–11 (2020). E6

23. Nowak, W., Nakayama, M., Kręcicki, T., Trypka, E., Andrzejak, A., Hachoł, A.: Analysis for extracted features of pupil light reflex to chromatic stimuli in Alzheimer's patients. EAI Endorsed Transact. Pervasive Health and Technol. **5**, 1–10 (2019). E4
24. Nowak, W., Nakayama, M., Trypka, E., Zarowska, A.: Classification of Alzheimer's disease patients using metric of oculo-motors. In: Proceedings of the Federated Conference on Computer Science and Information Systems (FedCSIS), pp. 403–407 (2021)
25. Pinkowski, B.: Robust fourier descriptions for characterizing amplitude-modulated waveform shapes. J. Acoust. Soc. Am. **95**(6), 3419–3423 (1994)
26. Ramsay, J., Hooker, G., Graves, S.: Functional Data Analysis with R and MATLAB. Springer, New York (2009)
27. Sivak, J.M.: The aging eye: common degenerative mechanisms between the Alzheimer's brain and retinal disease. Invest. Ophthalmol. Vis. Sci. **54**(1), 871–880 (2013)
28. Sørensen, H., Goldsmith, J., Sangalli, L.M.: An introduction with medical applications to functional data analysis. Stat. Med. **32**, 5222–5240 (2013)
29. Tombaugh, T., McDowell, I., Kristjansson, B., Hubley, A.: Mini-mental state examination (MMSE) and the modified MMSE (3MS): a psychometric comparison and normative data. Psychol. Assess. **8**(1), 48–59 (1996)
30. Zhang, D., Lu, G.: A comparative study on shape retrieval using fourier descriptors with different shape signatures. In: Proceedings of the 5th Asian Conference on Computer Vision, pp. 646–651. Springer, Berlin (2002)
31. Zivcevska, M., Blakeman, A., Lei, S., Goltz, H.C., Wong, A.M.F.: Binocular summation in postillumination pupil response driven by melanopsin-containing retinal ganglion cells. Vis. Neurosci. **59**, 4968–4977 (2018)

Chapter 6
Conclusions

6.1 Summary

Our aim was to develop diagnostic procedures using features of pupil reactions.

We believe that this approach is attractive because:

1. Pupil reactivity is the result of a complex anatomical-physiological mechanism, which may reflect the correctness of the structure and the function of the anatomical structures of the human body which are involved.
2. Measurement of the pupil's reactivity involves recording its image as a function of time. It is therefore a fairly simple, quick, and easily accessible measurement.

It seems, therefore, that the diagnostic procedure developed could provide a timely alarm indicator which can identify patients needing further, more sophisticated diagnostics.

Our method of analysis was based on the following approach:

1. A measurement experiment entitled "Specific pupillary reactivity in a given disease entity" was carried out on a selected study group, which included a control group and a patient group. Specific pupil reactivity is understood as the dynamics of pupil size in response to a specific stimulus or to spontaneous pupil reactivity.
2. Determination of the quantitative parameters that define the pupil dynamics of the study group
3. Use of advanced classification and analytical methods
4. Evaluation of the feasibility of using a specific approach as a screening technique

Chapter 1 analysed the possibility of using parameters that describe the shape of the pupil's response to light flash stimulation on patients with Age-Related Macular Degeneration. The aim of the analysis was to develop a diagnostic test that would distinguish between diseases affecting the retinal photoreceptors and

---

Chapter Co-author: Wioletta Nowak

M. Nakayama, W. Nowak, *Development of Diagnostic Procedures Using Features of Pupil Reactions*, Behaviormetrics: Quantitative Approaches to Human Behavior 21, https://doi.org/10.1007/978-981-95-4502-5_6

diseases affecting retinal ganglion cells. Research indicates that the activity of these photoreceptors and cells depends on the wavelength of the light entering the eye. In our study, chromatic light (red/blue) of two levels of luminance (low/high) was used to stimulate pupil response. Fourier descriptors were used to extract PLR waveforms. To detect patients affected with AMD, multidimensional scaling (MDS) and a clustering technique were used. To quantitatively assess detection performance, a set of combined features (patients, elderly and healthy subjects) was classified using the Random Forest method with optimisation using different levels of classification error. The results obtained, while identifying the potential of the method to discriminate between healthy AMD patients using chromatic PLRs, indicated the need to pay attention to the issue of patients with age-affected eyes in addition to this, as error rates for patients with age-affected eyes were not low.

Chapter 2 attempts to identify Alzheimer's patients using features of PLR response to light pulses. It was observed that the pupil size for AD patients is relatively small and responds slowly. In this study, monochromatic (red, blue) and polychromatic (white) light pulses with low/high levels of luminance were used to stimulate pupil response. Fifteen parameters were proposed to describe the shape of the pupil response, describing temporal, velocity, and acceleration characteristics. Two factor scores were extracted from the 15 features across all measurements, and logistic functions were introduced in order to calculate the probability of identifying AD patients. The use of additional analytical techniques, i.e. Bayesian technique and LASSO, allowed a higher level of prediction performance than using the previous procedure.

Two important conclusions can be drawn from the analyses obtained:

1. The contribution of the colour of the light, in particular blue light, is the key to accurate prediction.
2. PLRs for blue light which result mainly from ipRGC can be used to evaluate the factor of ageing independently.

Another approach is described in Chap. 3. Pupil-based ocular information (fluctuation over time of pupil size and pupil centre position) was recorded during a set of dementia tests of participants with Alzheimer's disease, mild cognitive impairment or in a control group. The hypothesis that dementia disorders reduce cognitive abilities was used here.

Pupil changes were assessed using frequency powers and some measurement characteristics of each eye. The Random Forest technique was used for the classification procedure. For the first time, a training data generation technique (data expansion technique) was used. A feature selection procedure (Boruta technique, Gini coefficients) was performed on the data in order to optimise prediction performance. The results obtained confirmed the contribution of fixation features and saccade information. Various disease entities manifested by pathological pupillary reactivity were selected for analysis. With some disease entities, attempts were made to include oculo-motors parameters in the analysis.

Chapter 4 proposes a procedure for assessing the level of dementia using the binocular pupil light reflex to chromatic light pulses on either eye. In this approach, the features of PLRs for blue and red light pulses of the left and right eyes were compared using PLR waveform factor scores, which are defined as the values of certain characteristic points of the PLR and its first and second derivatives. Based on factor scores the regression analysis was performed. It was confirmed that there are significant differences in the factor scores for blue and red stimuli. The analysis also confirmed the dominance of blue light. Unfortunately, the influence of a subject's level of dementia on PLR features was not confirmed during the above analysis. One of the possible reasons is the dependence on the accuracy of measurement, as feature extraction is based on point estimation. Another problem may be that the extracted features focus on the constriction phase of PLRs without allowing for the restoration phase which follows. It was therefore decided to add parameters describing the restoration phase and oculomotor parameters to the set of parameters analysed. The analysis was carried out using two logistic regression functions, and prediction accuracy was evaluated and made to conform using AUC (Area under the curve) metric.

In Chap. 5, for the description of pupil reactivity characteristics in response to chromatic light pulses in participants with differing levels of dementia, the use of coefficient vectors for B-spline functions as a technique of functional data analysis (FDA) technique was proposed. The data spaces of the extracted features of the two group of participants were trained using a sparse logistic regression function with Elastic-Net and Lasso technique, which is a variable selection technique. Classification performance was evaluated as a level of accuracy using the area under the curve (AUC) of ROC of the sampled data. The accuracy is around 70%.

6.2 Final Conclusions

The authors of this study analysed the use of different analytical approaches with different types of measured pupil dynamics in response to diagnose a number of human disorders manifested using pathological pupil reactivity.

Unfortunately, the resulting accuracy of the pathological function predictions is still insufficient for clinical application. Nevertheless, the analyses presented did allow some significant conclusions to be drawn. When analysing pupillometry study results, attention should be focused on the existence of several factors which significantly affect pupil reactivity characteristics and which may be divided into factors indirectly related to the disorder being analysed and factors completely unrelated to it. These are the age of the patient, baseline pupil size, the correctness for the measurement, the repeatability of the measurement conditions, and the procedure for the extraction of features.

In general, pupil size decreases with age, so age-related miosis and cataracts may influence pupil reactivity. As the factor of age may be the key to detecting the level of dementia, for example, a more detailed investigation should be considered instead of merely stating the actual age of the patient. It may be that changing the recording duration or stimulus parameters well, for example, increase the repeatability of the measurements and will have a direct impact on prediction accuracy.

A more robust procedure for feature extraction and prediction may require a more detailed set of features, which will also increase prediction accuracy. It should also be emphasised that a considerable limitation of the analyses performed is that all were carried out on different and often not very large study groups. The accuracy of patient classification is also an extremely important issue. Disease severity should be assessed objectively and evaluation needs to be conducted very carefully.

MIX
Papier aus verantwortungsvollen Quellen
Paper from responsible sources
FSC® C105338

If you have any concerns about our products,
you can contact us on
ProductSafety@springernature.com

In case Publisher is established outside the EU,
the EU authorized representative is:
Springer Nature Customer Service Center GmbH
Europaplatz 3, 69115 Heidelberg, Germany

Printed by Libri Plureos GmbH
in Hamburg, Germany